T0292380

Concise Guide to Computing Foundations

Kevin Brewer • Cathy Bareiss

Concise Guide to Computing Foundations

Core Concepts and Select Scientific Applications

Kevin Brewer
Olivet Nazarene University
Engineering
Bourbonnais, IL
USA

Cathy Bareiss
Olivet Nazarene University
Computer Science
Bourbonnais, IL
USA

ISBN 978-3-319-29952-5 ISBN 978-3-319-29954-9 (eBook)
DOI 10.1007/978-3-319-29954-9

Library of Congress Control Number: 2016951489

Printed on acid-free paper

This Springer imprint is published by Springer Nature
The registered company is Springer International Publishing AG Switzerland

Preface

The *Concise Guide to Computing Foundations* is designed to meet two main goals. The first goal is to help science students understand how the computers work so they can better use the computers and software when doing their work. The second goal is to provide an introduction into computational science. This book introduces topics from computer science that are important for scientists to understand. This includes (but not limited to) understanding how simulations work (including their strengths and limitations), understanding the computer's precision, and how much work the software needs to do to complete the task. It also takes these and other concepts in computer science and applies them to different topics in a number of scientific disciplines to help the science students understand how the two areas (science and computers) interact. The book is designed to be taught to second semester freshmen or sophomore science majors who have already had one major science lab course in their major and are ready to take calculus.

Why the Book Was Written

The book is the result of an NSF CCLI grant to study a new approach to teaching noncomputer science majors what they need to know about computing. Other introductory courses for nonmajors typically include one or more of the following: (a) how to program, (b) how to use office suite (and other software), (c) an overview of how computers work, and (d) ethics of computing and society. The need for the first two (programming and using office suite software) has become less essential for the following reasons. With the development of extremely powerful applications and simulations, fewer non-computer scientists need to program. The software needed to do advanced work often already exists and can be customized to the needs of the user without strong programming skills. In addition, many users are already able to use a computer for everyday tasks (via word processing, spreadsheets, email, browsing the web, etc.).

However, as society becomes more and more dependent on computers, users are trusting them more without understanding how they work. There are aspects of the working computing that are important for users to understand. These aspects vary from area to area. For scientists, it is important to understand the precision of computers and sensors so that they understand where errors can creep into their results. Musicians need to understand what is lost when audio files are compressed. Researchers that use computers to study language and literature need to understand how computers understand and interpret language. Every discipline depends on computing in an increasing amount. But it is essential for the user to understand what is happening within the computer to know how to use the results intelligently.

This textbook is an attempt to develop this new approach to computing for non-computer science majors. We decided to work in the area of the sciences because we were most familiar with that area and had a strong interest of some science faculty to help us. We hope this book can help the science students become better scientists by understanding how the computer helps to solve their problems.

Why You Should Be Interested in This Book

There are three primary reasons someone would be interested in this book. The first is that computing is everywhere and it is essential to understand many aspects of how a computer works. Today's society is extremely dependent on computers. These devices impact just about every aspect of life. The impact is not a surface impact any more. Computers are fundamental to almost every area. We rely on them to make our society work at the level it does. And as time progresses, this reliance will only become greater. It is very short sighted to develop a society where every aspect of it is very dependent on a technology that very few have any understanding of. Most of modern life is dependent on electricity. While most people may not understand the details of how electricity works, they know enough to know how to use it wisely, when it is dangerous, and what to do if it fails. But many people in today's society lack a similar level of understanding about computers. We don't understand what they are capable of, why they may not work, where the limitations are, and where errors can occur. This book seeks to address some of these areas for non-computing professionals. It aims to help the average person better understand important aspects of computing in the areas in which they work.

This leads to the second reason why someone would be interested in this book. This book offers a new approach on educating the general student about computing to address the needs of today's society. As mentioned previously, while knowing how to use office suite software is a valuable skill, it does not help the user understand how a computer works. While knowing how to program a computer does help users understand the workings of a computer, most users will not need to write programs. The knowledge that would be more useful is understanding how computers do their work at a level that impacts the results. How computers convert electrical signals to words on the screen is not important to most people (just as the details of how a battery works are not to important to quality use of electricity). However, scientists need to understand the limits of precision in a computer just as most people need to understand that it is dangerous to overload an electrical circuit.

This level of knowledge is what many of today's students need as they prepare for a future where computers are very pervasive.

The third reason for someone to be interested in this book is because they want to be better in their profession. As people use computers in careers, they can view the computer as an able assistant that they manage or they can view computers as a black box that they need to accommodate. Those that understand how the computer works in their profession can optimize the use of the computer to meet their needs. Those that don't have to adjust their work to the computer. The first type of person is much more productive in their job and able to do more difficult work. This book seeks to help develop such a person. While the focus on this book is for STEM professionals, the lessons can be applied (and expanded) to all other disciplines.

Who Is This Book for?

This book is aimed for STEM majors who have taken at least one lab science course in their major. These students should be excited about science, technology, engineering, and mathematics and feel comfortable in at least one of these areas. It assumes a basic understanding of the scientific method. There should be a strong motivation to study these areas. While the student is not required to do any calculus, there is some calculus shown in a few modules. Students cannot be fearful of mathematics and at least ready to study calculus. This book is also aimed at the sophomore level. This is to ensure that students are ready to do discovery learning (the primary method of the book) but also leaves them time to pursue additional work in this area later in their degree.

Additional Considerations

Those using this book should expect some additional benefits. Because this book takes examples from a number of different STEM areas, students will start to make connections between the different STEM branches. A student may experience a connection between chemistry and biology. Another might see a connection between geology and engineering. A third might see how mathematics and computing relate. There are many opportunities to build these bridges between all the different disciplines. Students will also get a good interdisciplinary introduction to computational science. This will spark enough interest in some of the students to pursue further studies in a very dynamic and promising field.

Organization of the Textbook

This book has two main categories of modules. This first category is computer science chapters. These chapters explain specific computer science concepts using different aspects of science as examples. These include modules on computational science, data representation, algorithms, self-defining data, and performance complexity. The second category is science modules where a specific science topic is explained and the connection to the related computer science is drawn. The areas of science include bioinformatics, chemical kinetics, engineering analysis, GIS, flow analysis, solving equations, curve fitting, optimization, and data acquisition. The book also includes a glossary and some short tutorials for some of the software used.

Guide to the Students

As is true with most of the things you study, you will get out of this what you put in. Try to engage the material as much as possible. There are "Discuss It" questions. Don't skip these. Discuss them with classmates. They are there to help you make important connections and to apply what you learn. Don't just try to complete the modules. Instead, always asked yourself what is this module trying to teach me at this moment about computing and the sciences. When working on modules that are not your discipline, try to make connections back to things you are familiar with. Try to apply the knowledge in other areas. To aid in doing this, try to work with students from other disciplines. This will broaden your knowledge and help you see things in a different way.

Guide to Instructors

You don't need to understand the details of all the STEM areas covered to be qualified to teach this course. This course has been taught by a computer science professor (with an interest in the sciences but not a trained scientist). This course has also been taught by a geology professor (with some background in computer science). The modules are designed to be self-teaching. So you can learn with your students. Teach them to how to be self-learners by demonstrating your own learning skills. Also, if the developers of this work (authors and contributors of given chapters) can be of any help, please contact us (contact information can be found at the book's website). All of us would like to be of help in any way possible.

The questions at the end of many of the modules (computing questions) are optional questions designed to help the students explicitly think about computing questions when working on modules in the sciences. Use them as you think appropriate.

If you are using this as the primary textbook for a course, it does not need to be covered sequentially (via the module numbers). We recommend that you use "just-in-time" methodology when picking the order to cover the modules. Cover enough of the computer science necessary for the other modules. Next is a dependency chart for each modules. Feel free to take any path through the modules. Also feel free to skip modules that you don't want to use.

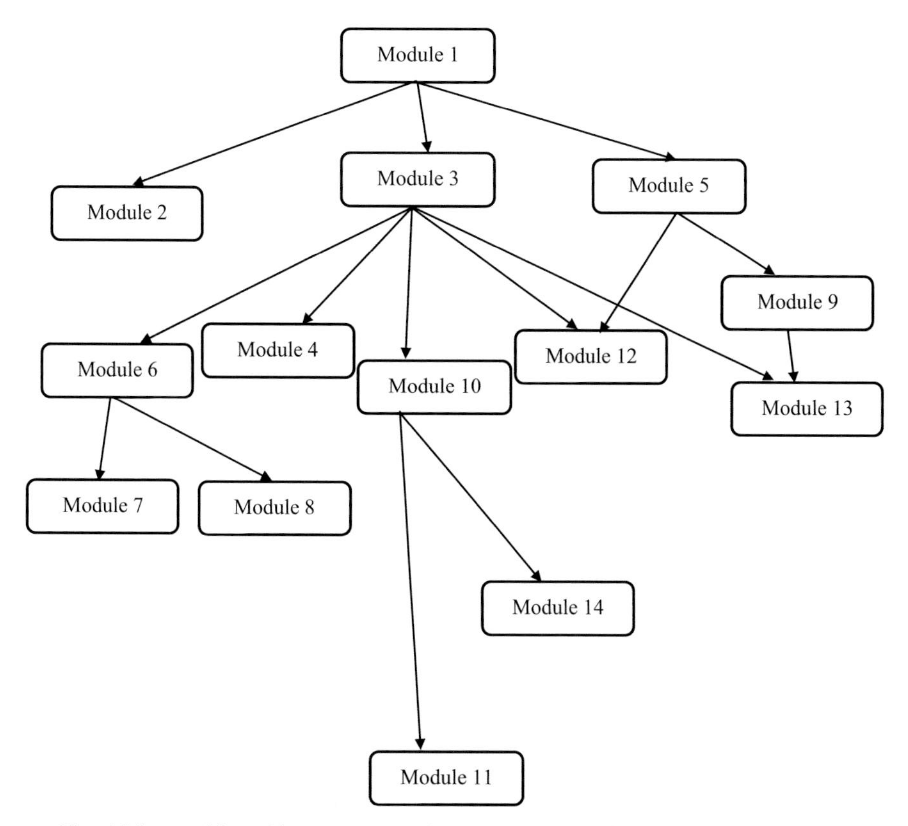

To aid in teaching this course, we have a number of resources. The first is an answer key. We also maintain on the website the software needed (and any extra files that would be helpful). The primary webpage for this book is http://www.springer.com/9783319299525. If you are interested in the background of this work and possible contributing additional modules to future additions, additional information can be found at: http://twiki.olivet.edu/twiki/bin/view/ComputationalScience/WebHome.

This work can also be used as supplements to other STEM courses. You can assign the students to work with the necessary computer science modules to do the modules associated with a specific course. You can also use material to supplement other courses by asking them specific computer science questions.

Uses of This as a Class

This work has been used as a class at the home institution for over 4 years. Each module assigned is allocated an average of a week. Students work in pairs going through the modules on their own with the instructor stopping the class every so often to ask important questions. There have been about two or three exams in the semester. The last 2 or 3 weeks of the class are devoted to semester projects. Students (working individually or in pairs) develop a project of their own choosing applying what they have learned. They then present the projects in class. It has been very insightful to see them apply things that they have learned to new areas. The students are also very excited to be given a chance to apply this work to something they are interested in.

Future of This Work

There are many ways this work can be continued. If you are interested in any of the following, please contact the authors. One area of expansion would be more STEM modules. A computer science module on data mining is needed. Other computer science modules might be very appropriate. In addition, other STEM areas might benefit from additional modules.

However, this new approach to teaching computer science to the general student can easily be expanded to non-STEM areas. Chapters on the foundations of computer science as used in music would be very appropriate. The same is true for all disciplines. If you are interested in expanding this work to new disciplines, please contact us.

Thanks

Many people were involved in this work. We would like to thank them all.

- Dr. Larry Vail contributed a great deal to this work (both in the writing of a number of the modules and in pioneering the teaching of this course).
- Dr. Willa Harper contributed the chemistry knowledge and was able to make sure that the computer science was not too technical.
- Dr. Greg Long contributed the biology knowledge and his excitement for the project keeps us going.
- Dr. Joe Schroeder helped with the engineering knowledge.
- The NSF was foundational in this work because they believe in the overall concept and funded the work.

Bourbonnais, IL, USA

Kevin Brewer
Cathy Bareiss

Contents

Introduction to Computational Science 1

1.1 Objectives

After completing this module, a student should be able to:

- Describe an example of a computational science simulation model.
- Define computational science, model, simulation, visualization, validation, verification.
- Appreciate the need to determine the reliability of simulation model results.
- List three sources of error in simulation model results.
- Appreciate the value of computational science.

1.2 Definitions

- Computational Science
- Model
- Bitmap
- Pixel
- Raster graphics
- Reliability
- Resolution
- Simulation
- Validation
- Vector graphics
- Verification
- Visualization

K. Brewer, C. Bareiss, *Concise Guide to Computing Foundations*,
DOI 10.1007/978-3-319-29954-9_1

1.3 Introductory Example

Before trying to define *computational science*, we will look at a NetLogo (Wilensky 1999) model of a rope in order to get a better feel for computational science. We want to open up the NetLogo model of a rope (Wilensky 1997a) using NetLogo Web: http://www.netlogoweb.org/launch#http://www.netlogoweb.org/assets/modelslib/Sample%20Models/Chemistry%20&%20Physics/Waves/Rope.nlogo

When going to the link, the top portion of the web page is the model interface. Below the interface are tabs: "Command Center" is where you can interact with the running model, or add code; "NetLogo Code" section contains all the simulation code; "Model Info" section is where you can read the information describing the model and how it works. Explore all the sections.

Notice the controls on the model. There are two buttons, *setup* and *go*, and there are three sliders: *friction*, *frequency*, and *amplitude*. There is also a timer speed slider at the top of the interface, and a *ticks* value displayed just above the model viewing window. All these controls change the behavior of the running model.

Push the *setup* button to initialize the model. Notice the red horizontal line that appears in the model window – this will be the simulated "rope". Your model is now ready to run. Push the *go* button to begin running the model. Observe the changing position of the "rope". Experiment with the *speed* slider to see how you can slow down or speed up simulation time. You can pause the model at any time by pushing the *go* button. Push it again to resume running. Initialize the model at any time by pushing the *setup* button. Continue to discover the behavior of the rope model by adjusting each of the *friction*, *frequency*, and *amplitude* slider controls. Observe the change in behavior of the rope as you change each slider (Fig. 1.1).

Answer It!

Perform an *experiment* and *observe* a corresponding change in the wave along the rope. The experimental variable will be *frequency*. Use values of 5, 10, 15, 20, and 25. For each input frequency, count the corresponding number of cycles in the wave along the rope. You will probably need to adjust the *speed* slider to slow down the simulation and push the *go* button to pause so you can accurately observe and record the dependent variable cycles.

Q01.01: Record your experimental data results in the table below.

Independent variable: Frequency	**5**	**10**	**15**	**20**	**25**
Dependent variable: Cycles					

Q01.02: Do you think the number of cycles is dependent on the frequency?

Q01.03: Give an English statement that describes the qualitative relationship between frequency and cycles.

Q01.04: Give a mathematical model (i.e. function) that describes the quantitative relationship between frequency and cycles.

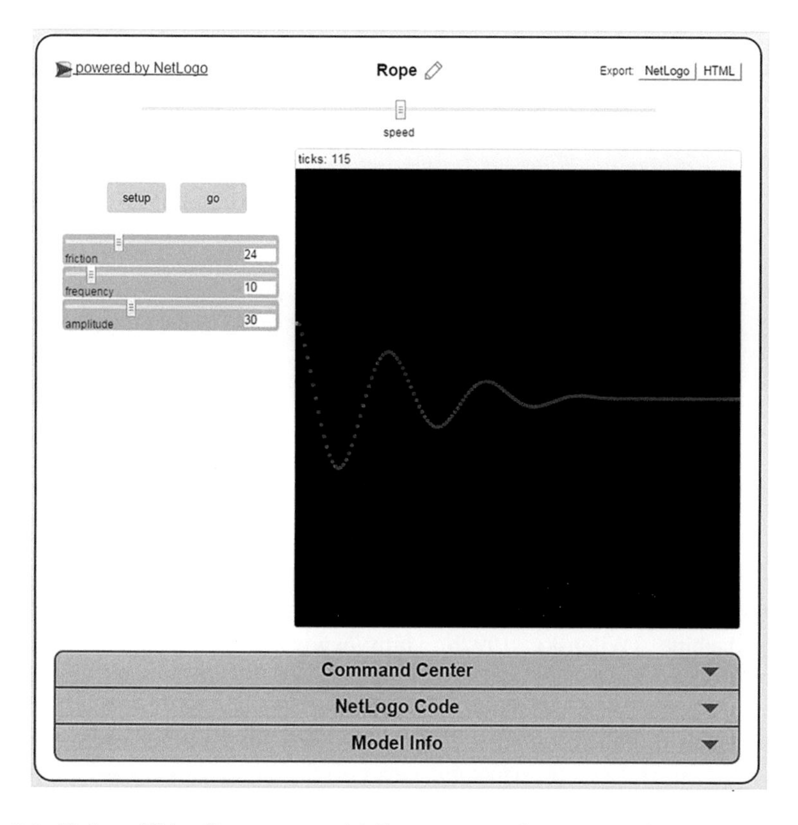

Fig. 1.1 NetLogo Web – Rope wave model (Screen capture from www.netlogoweb.org)

Continue experimenting by changing the *amplitude* input and observing the number of cycles. Then change the *friction* input and observe the number of cycles.

Q01.05: How does the wave change with a change in amplitude?
Q01.06: Do you think the number of cycles is dependent on the amplitude? Why?
Q01.07: How does the wave change with a change in friction?
Q01.08: Do you think the number of cycles is dependent on the friction? Why?

Notice how the waves often seem to travel from left to right. See if you can find inputs that make the wave *reflect* off the wall and travel back from right to left. You may also observe a place or places along the rope where the wave doesn't move up or down, left or right. It is just still. Place your finger or another pointer at a point on the rope to make sure the point is not moving. Change the *frequency* and watch for new waves *interfering* with each other until a new wave pattern is established.

Discuss It!
There are ways the simulation model could be changed by adding other inputs or changing the fixed assumptions in the model. For example, you could add a rope length input or allow the rope to be wiggled on both ends. What else could be done?

Answer It!
Q01.09: Briefly describe one change that could be made to the rope wave model to make it either more visually appealing, easier to control, or to make it more flexible to simulate other conditions.

1.4 Another Example

Before returning to the question, "What is computational science?" we will explore one more simulation model. Open up the web version of the NetLogo model of a 2D Wave Machine (Wilensky 1997b): http://www.netlogoweb.org/launch#http://www.netlogoweb.org/assets/modelslib/Sample%20Models/Chemistry%20&%20Physics/Waves/Wave%20Machine.nlogo

Experiment with the controls on this model. Look for *traveling waves* from the driver to the edges. Look for *reflecting waves* off the edges back toward the center. Look for *standing waves* on the surface. Look for *interfering waves* causing *turbulence*. Remember that you can slow down and pause the simulation (Fig. 1.2).

Discuss It!
How are the rope wave model and the wave machine model different? How are they similar?

Answer It!
In the wave machine model, change the *three-d?* switch to *off*.

Q01.10: With the *three-d?* switch off, how does the visualization show that the membrane surface is below the edges? Above the edges?
Q01.11: With the *three-d?* switch off, is the wave machine a 2-dimensional or 3-dimensional model?
Q01.12: Did the computational model change when the *three-d?* switch changed or did the visualization change?
Q01.13: Is it easier to "see" what is happening in the wave machine with the *three-d?* switch off or on? Why?

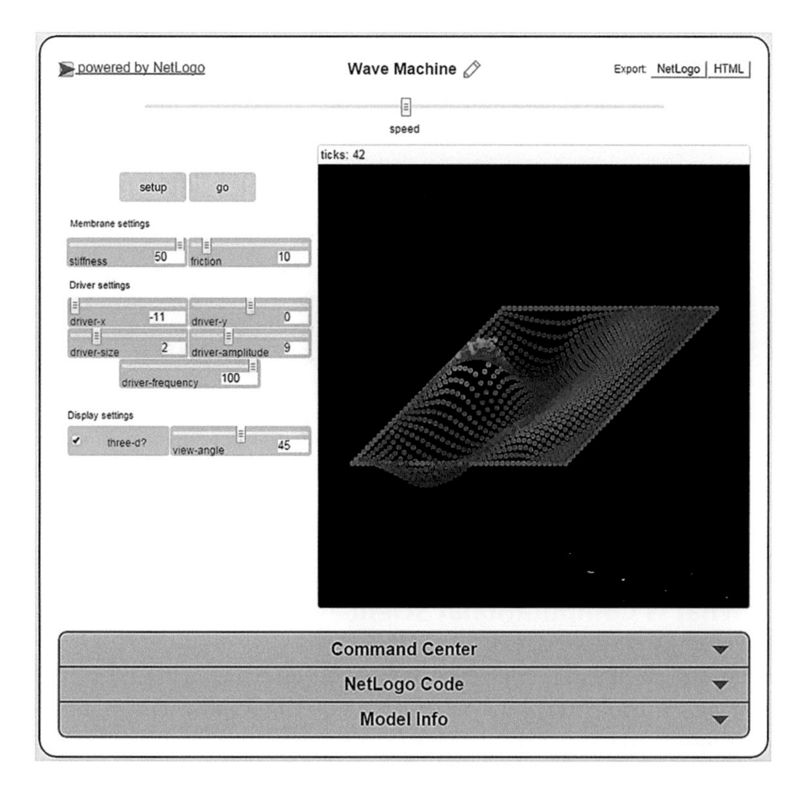

Fig. 1.2 NetLogo Web – Wave machine model (Screen capture from www.netlogoweb.org)

Discuss It!

Has experimenting with the NetLogo rope wave model taught you anything about real ropes and waves? What would make a computational model most useful?

Answer It!

Q01.14: Does the rope wave model behave like a real rope?

Q01.15: Is the rope wave model a "good" model?

Q01.16: How could you decide if the rope wave model is a "good" model?

You have not yet examined the details of how the rope wave model works. You cannot be certain if its internal model is close to a known mathematical model of waves in ropes. One source of error in a computational model is using the wrong or a poor mathematical model. Builders of a computational model may choose a good mathematical model and procedure (i.e. algorithm) for the science but not correctly translate them into the computing environment.

A second source of error in a computational model is incorrectly implementing the mathematical model. NetLogo, the Internet, and your computer make up a complex computing environment. Each of these components has limitations and could contain errors in computations. A third source of error in a computational model is inaccurate or incorrect computations in the computing environment.

Discuss It!
How can you be confident in using the results from experiments performed with computer simulation models?

Answer It!

Q01.17: List and rank (in your opinion) the three sources of error in simulation model results for the wave and rope models you just used.

1.5 What Is Computational Science?

Let's return now to the initial question, "What is computational science?" On the surface, *computational science* must involve computation and science – so let's first explore "science".

Traditional science advances when someone (1) carefully observes something, (2) develops a hypothesis or theory, and (3) designs experiments and tests the hypothesis or theory. When the experimental data supports the hypothesis or theory, mathematical models are developed that can be used to predict future outcomes.

There are often drawbacks to traditional science. As mathematical models become more complex, the algorithmic procedures required are increasingly time consuming, tedious, and error-prone. Modern computer systems can perform algorithms quickly, tirelessly, and without computational error.

Computational science is therefore the intersection between science, math, and computing. Figure 1.3 shows a way of describing computational science.

Two legs of the triangle shown in Fig. 1.3, *Theory* and *Experiment*, are covered by the traditional scientific method. Experimental data is collected using observational tools such as human eyes, photography, microscopes, telescopes, and measuring devices (e.g. stopwatches, rulers, micrometers, balances, and graduated cylinders). Two of the vertices, *Application* and *Algorithm*, are also included in traditional science. The mathematical model must be validated by its ability to usefully predict events in the real world.

The third leg, *Computation*, can be used to support any application domain of science (e.g. computational physics, computational chemistry, computational biology, computational engineering, and computational geology). The computer can even be used in the study of computer science. Computational science involves all aspects of traditional science within the bounds and limitations of today's computing environments.

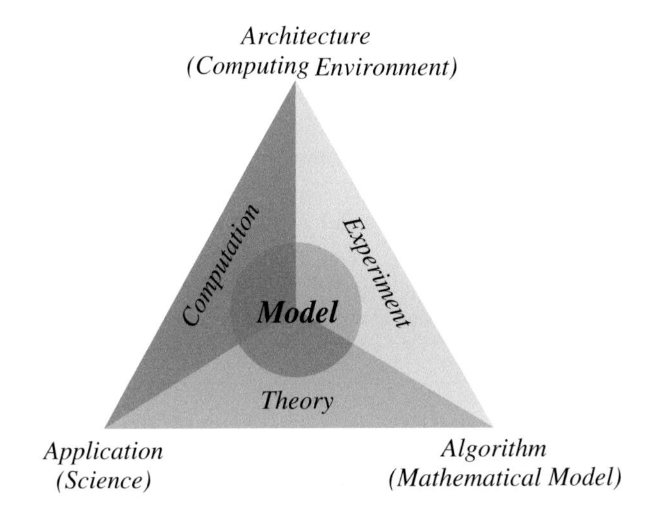

Fig. 1.3 Overview of computational science (Image from www.shodor.org)

Let's now explore our previous Rope Wave simulation to see the mathematical model and computational algorithm that were used to simulate a rope.

Again open up the NetLogo model of a rope using NetLogo Web. Select the "NetLogo Code" section at the bottom of the page. Don't worry right now about understanding exactly what the program means, but you have now seen a computer model. As the computer program runs in the NetLogo environment, it performs a computation that simulates the model over time and shows a visualization of the computation for the user. The remaining modules in the book will explore the computing foundations for this and other science computer simulations.

Discuss It!

Performing a real world experiment and collecting data often raises difficult issues such as measurement, time, cost, and ethics. Think of some examples of experiments with these issues. In these situations, computational science models should be considered because they may be very useful.

Answer It!

Q01.18: Describe a real world experiment with a *measurement* issue. How might a computer simulated experiment alleviate the *measurement* issue?

Q01.19: Describe a real world experiment with a *time* issue. How might a computer simulated experiment alleviate the *time* issue?

Q01.20: Describe a real world experiment with a *cost* issue. How might a computer simulated experiment alleviate the *cost* issue?

Q01.21: Describe a real world experiment with an *ethical* issue. How might a computer simulated experiment alleviate the *ethical* issue?

1.6 Related Modules

- Module 2: Chemical Kinetics. Simulation types are introduced.
- Module 3: Data Representation, Abstraction, Limitations. Computational limits/errors (e.g. round-off, overflow, underflow, limited precision, non-reproducible computations) are examined further.
- Module 5: Procedures: Algorithms and Abstraction. Algorithms and computation are examined further.

Acknowledgement The original version of this module was developed by Dr. Larry Vail.

References

Shodor.org. What is Computational Science? http://www.shodor.org/refdesk/Help/whatiscs. Retrieved June 2011

Wilensky U (1997a) NetLogo Rope model. http://ccl.northwestern.edu/netlogo/models/Rope. Center for Connected Learning and Computer-Based Modeling, Northwestern University, Evanston

Wilensky U (1997b) NetLogo Wave Machine model. http://ccl.northwestern.edu/netlogo/models/WaveMachine. Center for Connected Learning and Computer-Based Modeling, Northwestern University, Evanston

Wilensky U (1999) NetLogo. http://ccl.northwestern.edu/netlogo/. Center for Connected Learning and Computer-Based Modeling, Northwestern University, Evanston

Types of Visualization and Modeling 2

2.1 Objectives

After completing this module, a student should be able to:

- Describe the conditions under which computer models may not be completely reliable.
- Compare and contrast two different types of modeling: agent-based and dynamic systems modeling.

2.2 Definitions

- Agent-based modeling
- Bimolecular
- Concentration
- Dynamic systems modeling
- Equilibrium
- Kinetics
- Rates
- Reaction
- Unimolecular

2.3 Motivation

Often in science, we investigate and explore processes that change over time. This module will introduce two classes of computational science for how you can use computers to model these processes: *agent-based* and *dynamic systems*. We will use the area of chemical reaction kinetics to explore these two classes.

K. Brewer, C. Bareiss, *Concise Guide to Computing Foundations*,
DOI 10.1007/978-3-319-29954-9_2

First, we will work with two models from NetLogo, a program that uses agent-based modeling. Agent-based modeling simulates autonomous agents that act and are acted upon, to view the impact on the whole system. In our examples, NetLogo will show a graph of the change in a reaction over time along with visual representations of individual molecules and their changes and interactions during the course of the reaction.

Next, we will investigate two models with the Vensim Personal Learning Edition (PLE) software, a program that utilizes dynamic systems modeling. Vensim allows the user to model a chemical reaction from the perspective of the chemical equations and the mathematical model of the reaction dynamics. The model will provide predictive information about the change in a reaction over time.

2.4 Introduction

Before we delve into the models, a brief review of chemical reaction kinetics is needed. Say we understand that when we combine two molecules in a beaker, a reaction will take place as described by the following chemical equation:

$$A + B \leftrightarrow C \tag{2.1}$$

where A, B, and C represent three different chemical molecules and the double arrow represents that the equation is reversible – that is, the chemical reaction can go both ways, forward from A + B to C, and in reverse from C to A + B. By convention, the left side components are called the reactants, with the right side components called the products. The reaction will reach a dynamic equilibrium, meaning there will be a stable number of A, B, and C molecules all in the beaker at the same time. (That does not mean the reactions cease, just that there is a balance between C being produced and being used.)

Note that Eq. (2.1) does not provide any information about the rate at which the reaction will take place, nor what the equilibrium conditions may be. To understand that, a reaction rate needs to be determined. The following shows a hypothetical relationship between concentration and time that could have been determined from an experiment (by putting the reactants A and B together in a beaker and measuring the concentration of A over time as the reaction takes place).

The since the ln[A] plot in Fig. 2.1 shows a straight line (linear) relationship, we would call this a first-order reaction. The slope of that line would be equal to ***k***, the reaction rate coefficient. Mathematically:

$$Rate = -\frac{d[A]}{dt} = k[A] \tag{2.2}$$

$$[A] = [A]_o e^{-kt} \tag{2.3}$$

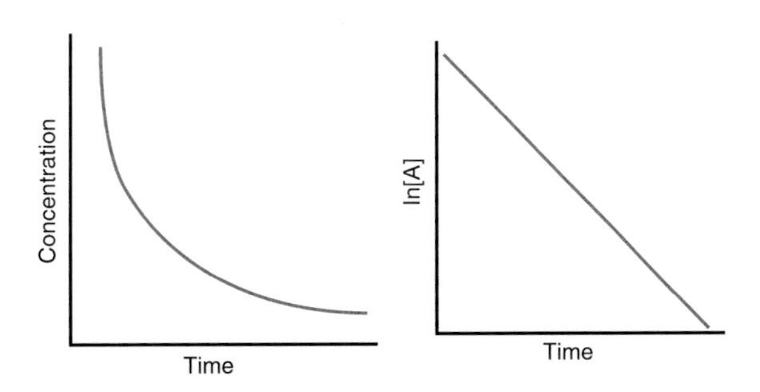

Fig. 2.1 Reactant concentrations vs time (From http://chemwiki.ucdavis.edu/)

where [A] is the concentration at some time, and $[A]_o$ is the initial concentration. Equation (2.3) is considered the rate law for the reaction going in the forward direction. A similar rate law could be experimentally determined for the reverse reaction.

The forward and reverse reaction rate coefficients can then be used in models to represent the chemical reaction over time.

Answer It!

Q02.01: What is a reversible reaction?
Q02.02: What is a reaction rate?
Q02.03: What occurs at equilibrium in a reversible reaction?

2.5 Agent-Based Chemical Kinetics

Open up the NetLogo model of Simple Kinetics 1 using NetLogo Web which can be found at http://www.netlogoweb.org/launch#http://www.netlogoweb.org/assets/modelslib/Sample%20Models/Chemistry%20&%20Physics/Chemical%20Reactions/Simple%20Kinetics%201.nlogo

The *Model Info* tab contains the description of the simulation and provides some basic information about the kinetics of a simple reversible reaction.

Choose the values of the rate constants, k_u and k_b with appropriate sliders. Rate constant k_b controls the rate of the reaction by which two green molecules turn bimolecularly into a single red molecule. Similarly the constant k_u controls the rate of the reaction, by which a red molecule turns unimolecularly into two green molecules.

You also need to set the initial number of molecules (*number* slider). You are allowed to choose anything from 0 to 100. Start with a reasonable number of around 100.

Having chosen values for the rate constants and initial molecules, click *SETUP* to clear the world and create an initial number of green molecules. (Note: this model only starts with green molecules, but one could program it to also start with some red molecules.) Click *Go* to start the simulation. You will see molecules floating

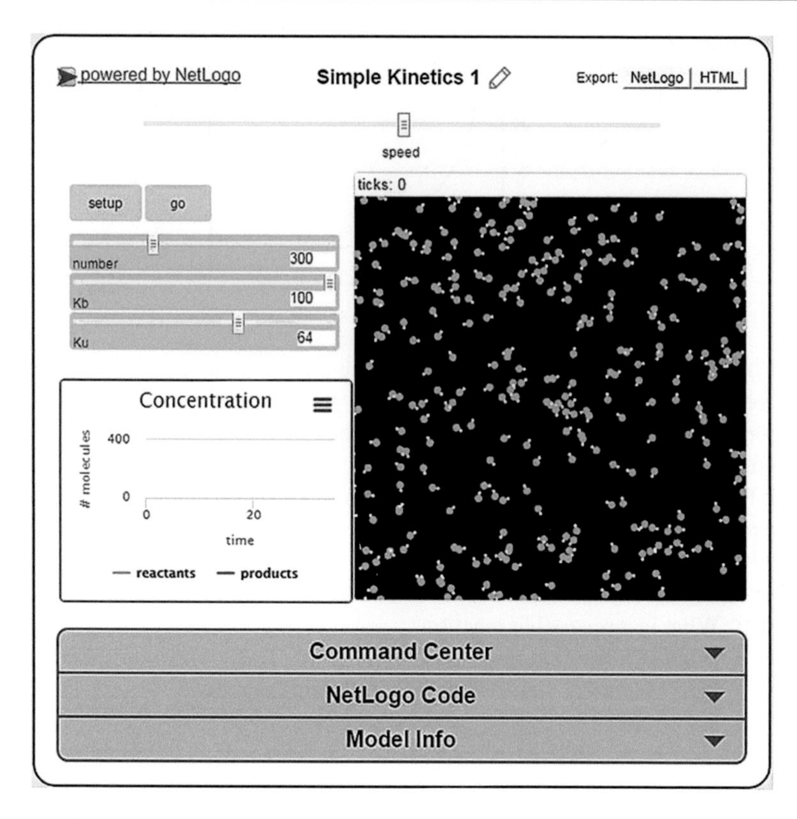

Fig. 2.2 NetLogo Web – Simple Kinetics 1 model (Screen capture from www.netlogoweb.org)

around the box and changing color. Pay attention to the plot of the concentrations. You can press *Go* a second time to stop the simulation. When you have seen what happens, try changing the sliders to different values, press *setup* and *go*, and observe what happens (Fig. 2.2).

Answer It!

Q02.04: What is the chemical reaction equation for this model? (In a form similar to Eq. (2.1).)

Q02.05: What variable name does the model use for the forward reaction rate coefficient?

Q02.06: What variable name does the model use for the reverse reaction rate coefficient?

Q02.07: What is the Principle of Stationary Concentrations? (See the Model Info tab.)

Q02.08: Do the plots soon reach stationary concentrations when running the model?

Q02.09: How does changing the concentrations of reactants and products (numbers of molecules) in the system affect the equilibrium?

Q02.10: Does it take more or less time to reach a stationary state under various conditions?

Q02.11: How do the stationary concentrations depend on the values of k_u and k_b?

Q02.12: You can change k_u and k_b while the model is running. See if you can predict what the stationary concentrations will be with various combinations of k_u and k_b. What were some of your predictions, and did you get what you expected?

Q02.13: What happens if you choose k_u, k_b, or both to be 0?

The way the Simple Kinetics 1 model is currently designed, it is a reasonable simulation for behavior of chemical reactions. However, can we always depend on our computer models? Try selecting *1* for the number of molecules (press *Setup* after you do this), then run the simulation. What happens? What happens to the molecules and the graph? What does this tell us about the reliability of computer simulations?

Now open up the NetLogo model of Simple Kinetics 2 using NetLogo Web: http://www.netlogoweb.org/launch#http://www.netlogoweb.org/assets/modelslib/Sample%20Models/Chemistry%20&%20Physics/Chemical%20Reactions/Simple%20Kinetics%202.nlogo This is a model that is very similar to Simple Kinetics 1, but with some important differences/additions as we shall discuss. As before, choose the values of k_u and k_b with appropriate sliders. You will not be able to choose as large a number for k_u as you did for the previous simulation.

What separates Simple Kinetics 2 from Simple Kinetics 1 are the features that have been added to simulate experimental changes other than rate constants: initial numbers of molecules, temperature of the reaction, and volume (the reaction box is shown with yellow walls that can vary in thickness).

Temperature changes have a unique effect on equilibrium compared with the other variables. You can observe this effect by toggling the *temp-effect* button on or off and using the slider to set the temperature of the reaction in Celsius. The initial number of molecules and the volume of the reaction also affects the concentration of the reactants and products (Fig. 2.3).

Start with the default *edge-size* and be sure that *temp-effect* is set to *off*. Click *SETUP* to clear the world and create an initial number of green molecules. Click *GO* to start the simulation. You should observe similar results to those you observed with Simple Kinetics 1.

Run the simulation several times, using the sliders and buttons to observe how initial numbers of molecules and reaction volume vary the concentration and affect the equilibrium. Then use the slider to observe how temperature affects the equilibrium. Finally, vary both concentration and temperature to see if your observed effects are impacted by some sort of interaction.

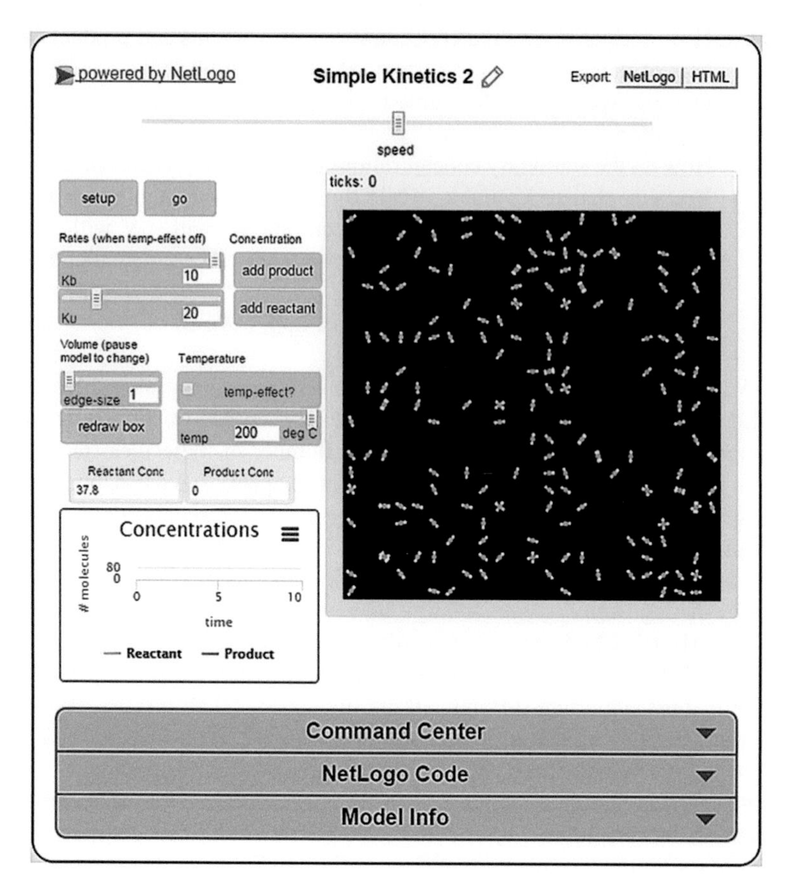

Fig. 2.3 NetLogo Web – Simple Kinetics 2 model (Screen capture from www.netlogoweb.org)

Answer It!

Q02.14: If you could alter this model so that you can simulate what happens to the reaction at 1000° Celsius, do you think it would make much difference to the time at which equilibrium is achieved? Why? (Hint: Extrapolate your observations from increasing the temperature.) Why might the result not be consistent with your observations with the current model?

Q02.15: Click on the *NetLogo Code* tab and find the portion of the program where the temperature effect is simulated. What equation is used to calculate this effect?

Q02.16: When reading the information in the paragraph about the temperature effect calculation, can you see something that indicates that there may be a difference between the model and a real-world experiment?

Q02.17: What does your above observations mean for the reliability of this computer model? What might we need to be aware of when dealing with computer simulations?

Discuss It!
At an atomic level, chemical reactions take place during collisions of molecules. From your experimenting and examination of the code for the NetLogo models you just used, do you think the models are simulating the reactions from collisions? If not, why do you think the molecules were shown during the simulation? Would it be possible (or practical) to simulate chemical reactions at the atomic collision level?

2.6 Systems Dynamics Chemical Kinetics

The systems dynamics approach to modeling is based on a feedback loop or causal loop of organizing a system. The Vensim Personal Learning Edition (PLE) software is an icon-based systems dynamics modeling package that is available for free (http://www.Vensim.com). It uses a tank and flow representation of causal loops which can be used to model a chemical reaction based on mathematical equations. This model can provide predictive information about the change in a reaction over time.

Recall our original first-order chemical reaction Eq. (2.2), slightly rewritten:

$$[A_{t_2}] = [A_{t_1}](1 - k\Delta t) \tag{2.4}$$

Therefore we can compute the concentration at time $= t_2$ if we know the rate coefficient k, the concentration [A] at the initial time (t_1). Similarly, for the next time, t_3, we need to know the concentration and time at t_2. This type of equation is a feedback loop since we need to feed the answer at t_2 back into the equation to get our answer for t_3, and so on.

2.6.1 Simple First Order Reaction

Consider a first order reaction similar to the one you worked with in Netlogo. A first order reaction is one in which there is a linear relationship between the concentration of single reactant and the rate of reaction. The following are examples of first order reactions.

$$H_2O_2 \rightarrow H_2O + \tfrac{1}{2}O_2 \tag{2.5}$$

$$2N_2O_5 \rightarrow 4NO_2 + O_2 \tag{2.6}$$

You can see from the above reactions that the number of molecules on the reactant side of the equation (one in Eq. (2.5), two in Eq. (2.6)) does not necessarily indicate the order of the reaction, A simple first order reaction can be indicated with the following simple chemical equation.

$$A \xrightarrow{k_1} B \tag{2.7}$$

One can use differential rate equations to describe the progress of this reaction toward completion as the concentration of A, [A] decreased, and the concentration of B, [B] increases. The rate at which this reaction occurs can be expressed as the change in concentration (quantity) of each substance in the reaction with respect to time.

$$dA/dt = -k_1A \qquad \underline{dB}/dt = k_1A \tag{2.8}$$

To model a reaction with Vensim, we must build the system and program the equations. For a first order reaction, we will use tanks to represent the concentrations of our reactants and products, and a flow "valve" to represent the rate of change (Eq. 2.8). The resulting Vensim model is shown in Figs 2.4, 2.5 and 2.6.

Answer It!

Create the first order chemical reaction simulation in Vensim. Use 1000 for the initial value of Reactant A, 0 for the initial value of Reactant B, and a value of 1 for k1. The Chemical Reaction equation will be the multiplication of k1 and Reactant A. The time step should be .0625, with a Final Time of 10 (seconds). Use the RK4 integration type. Graph the values of Reactant A and Product B.

Q02.18: What is the sum of *A* and *B* at any given time?

Q02.19: Thinking back to the original reaction that you are modeling and the starting conditions for the chemicals (A and B), can you explain why the sum of A and B at any given time equals this value?

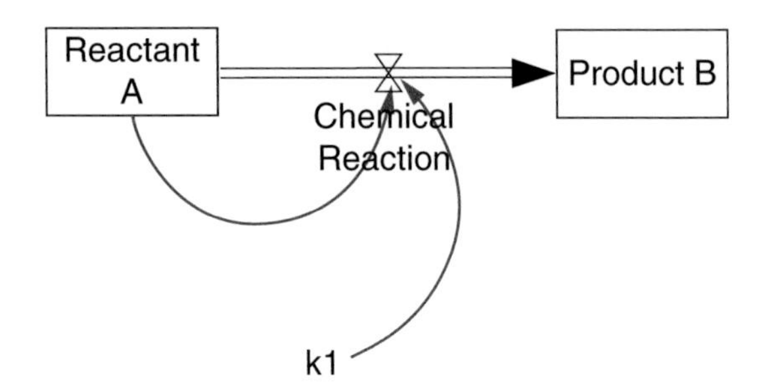

Fig. 2.4 First order reaction in Vensim

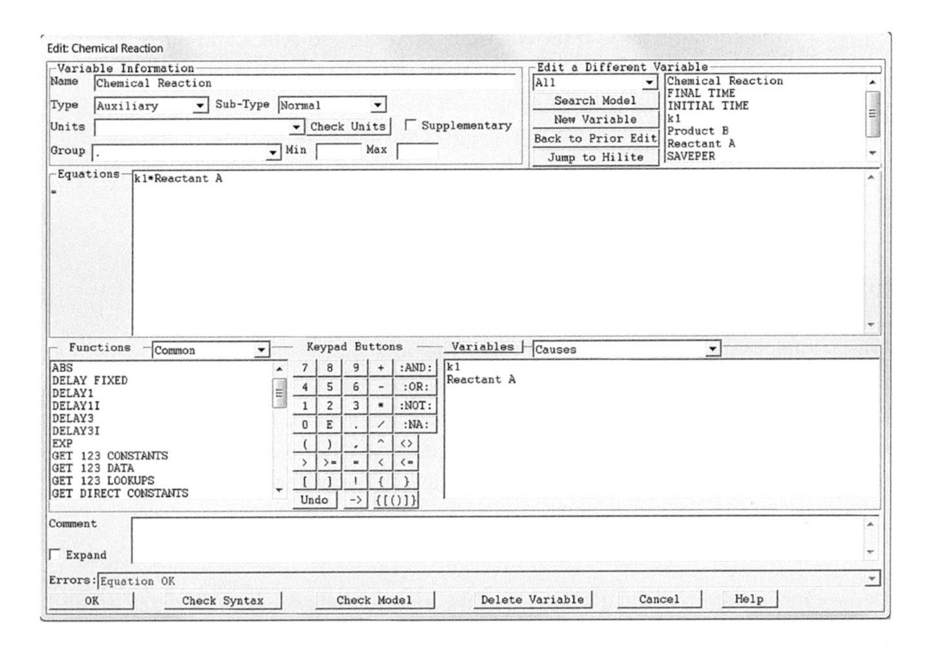

Fig. 2.5 Reaction rate equations in Vensim

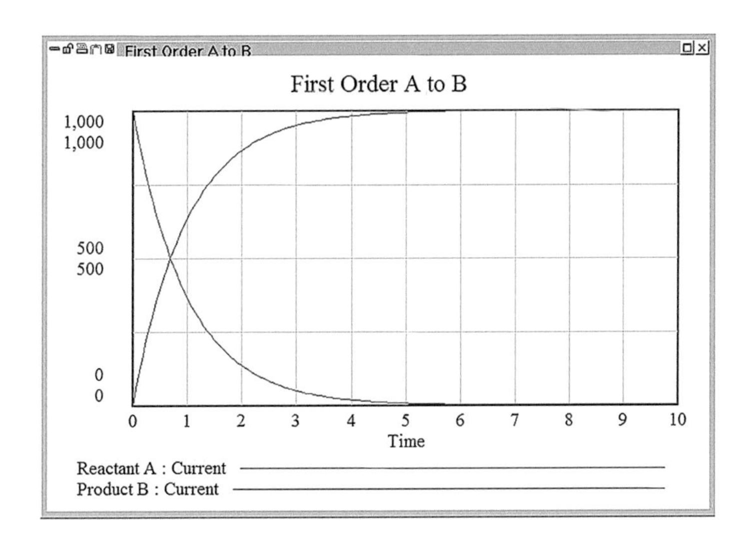

Fig. 2.6 Example output from first order reaction simulation

Q02.20: At what approximate time does A = B (50 % completion)? Why might this point in the reaction be important for this reaction?

Q02.21: How would the graph change if $k_1 = 2$? You can check your answer by clicking the *Equation* button. Click on k_1 in the model. To the right of the equation sign in the dialogue box, change *1* to *2* and click *OK*. Run the model again and display the graph.

Q02.22: How does the graph change? Is this consistent with what you would expect, given the equation that you are using for the rate of this reaction?

2.6.2 Reversible First Order Reactions

A reversible reaction is one in which the reaction moves in both the forward and reverse directions. It is designated with a double-headed arrow and indicates an equilibrium (see definitions). A reversible first order reaction is one in which both the forward and reverse directions are first order. An example reaction would be the dissociation of a proton from a weak acid such as hydrofluoric acid.

$$HF \rightleftharpoons H^{+} + F^{-} \tag{2.9}$$

As mentioned before in the discussion of the simple first order reaction, the number of particles on either side of the equation does not necessarily indicate the order of the reaction.

The following equation is the simplest general representation of a reversible first order reaction.

$$A \underset{k_2}{\overset{k_1}{\rightleftharpoons}} B \tag{2.10}$$

The differential rate equations are:

$$dA/dt = -k_1A + k_2B \qquad dB/dt = k_1A - k_2B \tag{2.11}$$

Answer It!

Build the model as shown in Fig. 2.7 by using the instructions that are given previously. Set $k_1 = k_2 = 1$ and the initial values of $A = 1000$ and $B = 0$. Be sure that the formulas you enter correspond to the differential rate equations. Run the model and set up an appropriate graph to see how A and B change with time

Q02.23: What is the sum of A and B at any given time?

Q02.24: How will the graph change if $k_1 = 2$ and $k_2 = 1$? Change the model and see.

Q02.25: Once equilibrium is established, is the ratio B/A the same as k_1/k_2?

Q02.26: Try several different numbers for A, B, k_1, and k_2. Were you able to find values that gave graphical results that seem incorrect compared those that you saw previously? If so, what were those values? Under what conditions do you think this type of model would be unreliable?

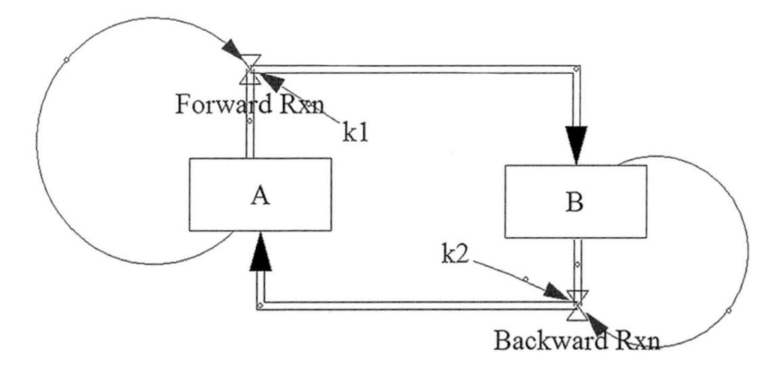

Fig. 2.7 Reversible first order reaction simulation in Vensim

Discuss It!
Which of the two classes of simulations (agent based or system dynamics) did you find easiest to implement to represent a first order chemical reaction? Which one was (potentially) more flexible to accommodate additional controls?

2.7 Computing Questions

- How reliable were the simulations you used/developed? Why?
- How valid where the simulations you used/developed? Why?
- How did you verify the simulations you used/developed?
- What were the possible sources are errors in your simulation(s)?
- Did your simulation(s) alleviate measure/time/cost and/or ethical issues? Is so, how?

2.8 Related Modules

- Module 1: Introduction to Computational Science. Simulations are introduced.

Acknowledgement The original version of this module was developed by Dr. Willa Harper.

References

Metz C (2011) Lab 10b. Computational Study of Chemical Kinetics Using Vensim PLE (GIDES). From Computational Chemistry for Educators, The Shodor Education Foundation, Inc.

Stieff M, Wilensky U (2001) NetLogo Simple Kinetics 2 model. http://ccl.northwestern.edu/netlogo/models/SimpleKinetics2. Center for Connected Learning and Computer-Based Modeling, Northwestern University, Evanston

Wilensky U (1998) NetLogo Simple Kinetics 1 model. http://ccl.northwestern.edu/netlogo/models/SimpleKinetics1. Center for Connected Learning and Computer-Based Modeling, Northwestern University, Evanston

Wilensky U (1999) NetLogo. http://ccl.northwestern.edu/netlogo/. Center for Connected Learning and Computer-Based Modeling, Northwestern University, Evanston

http://phet.colorado.edu/en/simulation/reactions-and-rates

http://www.science.uwaterloo.ca/~cchieh/cact/c123/chmkntcs.html

3 Data Types: Representation, Abstraction, Limitations

3.1 Objectives

After completing this module, a student should be able to:

- Define the different basic data types (i.e. integer, floating point, and character) and the implications and possible errors of each.
- Define different sources of error (i.e. measurement, representation, round off, overflow, underflow, and interpretation).
- Appreciate the issues associated with data storage (i.e. including space requirements, magnitude of size, compression).
- Apply abstraction to data in the areas of trees, arrays, linked lists, and graphs.

3.2 Definitions

- Abstraction
- Array
- Binary tree
- Data structure
- Floating point
- Graph
- Integer
- Linked list
- Tree

K. Brewer, C. Bareiss, *Concise Guide to Computing Foundations*,
DOI 10.1007/978-3-319-29954-9_3

3.3 Motivation

The foundation of science and engineering is data. We measure and observe nature and experiments and collect data (typically numbers, but could be other types of information). We analyze and process that data to arrive at our conclusions and we base actions on our interpretation of data. Increasingly in our world, we use computing devices to collect, record, store, analyze, report, and share our science and engineering data. Every scientist and engineer needs to understand how the computer stores data.

3.4 Abstraction

A key theme throughout this module will be *abstraction* and we will explore five levels used in storing data and information. The computer stores everything as electrical signals. Our first abstraction is to take the electrical signals into 0's and 1's (Abstraction 1). From that we use 0's and 1's to store integers (Abstraction 2). We then use integers to store floating point numbers (Abstraction 3) and characters (Abstraction 4). We finally can combine those three items (integers, floating point numbers, and characters) to store facts and information in different ways (Abstraction 5).

In conjunction with abstraction and storing different types of information, we will also consider limitations that the computer places on these methods and ways to overcome these limitations. In addition, we will to consider timing and space issues associated with each method.

Abstraction 1: Electrical Signals to 0's and 1's
Currently, all computers are based on electrical signals (with some optical signals used between computers). While there are ideas about using other forms such as DNA (Sample and correspondent 2007) and organic molecules ("Scientists Create Organic 'Molecular Computer'" 2016), these are primarily only in the research stage. Thus we will look only at electrical computers.

Further, when dealing with electricity and the computers, the standard way is to think of it in one of two stages: *on* or *off*. But, just like a ceiling fan can have four different speeds (off, slow, medium, fast), we are not limited to on/off when working with computers. Some computers have been made/proposed that have more than two values (Neeley et al. 2009), but all modern computers are based solely on the on/off principle (called *binary*) – which will be the focus of this module.

Because of the inconsistency of electricity at low powers and the rapid changes used in computing, we can't use just on and off. Instead, one range of voltage will be used for off and a second range will be used for on. One common standard is for off (0) to be between 0 and 0.4 volts and on (1) to be between 2.4 and 5 volts. Other standards exist for different uses, but each will use a range with varying voltage gaps to ensure reliable differentiation between 0's and 1's.

Abstraction 2: 0's and 1's to Integers

While a *number* is a concept, a *numeral* is a representation of that concept. Numbers don't vary (e.g. if there are five blocks, there are five blocks). However, there are many different ways to represent those five blocks. We can say that there are *5 blocks*, or that there are *V blocks*. We could even use different colors to represent each number. A numeral is just a way to communicate a number. One category of numerals uses different bases. Most people think naturally in base 10 (because we have 10 fingers) which means we use only ten different single symbols before we add additional symbols to represent larger numbers. For example, to represent eight in base 10, we use the symbol "8". To represent the number twelve, we use two symbols, "12".

So to represent six hundred and seventy five in base 10, we use:

$$675_{10} = 6*10^2 + 7*10^1 + 5*10^0 = 6*100 + 7*10 + 5 \tag{3.1}$$

But if we used a different base, the "675" representation would mean something different:

$$675_8 = 6*8^2 + 7*8^1 + 5*8^0 = 6*64 + 7*8 + 5 = 445_{10} \tag{3.2}$$

So for base 8, we would represent six hundred and seventy five as:

$$1243_8 = 1*8^3 + 2*8^2 + 4*8^1 + 3*8^0 = 1*512 + 2*64 + 4*8 + 3 = 675_{10} \tag{3.3}$$

And so "2143" symbols in base 5 are:

$$2143_5 = 2*5^3 + 1*5^2 + 4*5^1 + 3*5^0 = 298_{10} \tag{3.4}$$

A base 2 example:

$$\begin{aligned} 101101001_2 &= 1*2^8 + 0*2^7 + 1*2^6 + 1*2^5 + 0*2^4 + 1*2^3 + 0*2^2 + 0*2^1 + 1*2^0 \\ &= 361_{10} \end{aligned} \tag{3.5}$$

Any base follows the same formula: (in base x) $\mathbf{abcd_x = a*x^3 + b*x^2 + c*x^1 + d*x^0}$

Answer It!

Q03.01: What symbols are used to represent numbers in base 10? (Hint: there are ten of them)

Q03.02: What symbols are used to represent numbers in base 4? In base 2?

Launch the Windows calculator and set it to *programmer mode* to answer the following:

Q03.03: What is 57 in binary? (i.e. base 2)
Q03.04: What is 1011010111101 in decimal?

Discuss It!
It was likely somewhat annoying to write out the details for that last two question because of the difficulty in keeping track of all the individual symbols in the proper position. One common way of solving that problem is to write a binary number in groups of 4 digits (starting at the right). What is 2^4? Why might computer professionals like base 16?

Q03.05: In base 16, what characters would be used to represent 10 through 15?
Q03.06: Computer scientists do a lot of work in base 16, but other larger bases are used in other situations (sometimes without people realizing it). What symbols might be used to represent base 36?
Q03.07: A car's VIN is 17 places long. Assuming a base of 36, how many different cars can exist until we run out of VIN's?

3.5 Limitations and Space Issues

Just as there are a limited number of possible license plate combinations, computers store data with a limited number of bits. For now, we will explore the limitations of representing integers in computers, and in dealing with the amount of space required to store the values.

When talking about a large amount of space, an adapted metric scale is used. The prefixes (kilo, mega, etc.) use the powers of 2 that are closest to the related powers of 10. These are found in Chart 3.1.

In computing, we will find that most improvements involve some tradeoff: benefits in one area means a cost in another area. For example, there must be a cost (increase in need of storage space, increase in time, etc.) when we allow for larger numbers.

Chart 3.1 Values for common prefixes

Prefix	Kilo	Mega	Giga	Tera	Peta
Power of 10	10^3	10^6	10^9	10^{12}	10^{15}
Power of 2 Value	1024	1,048,576	1,073,741,824	1,099,511,627,776	1,125,899,906,842,624

Answer It!

For this activity we will use the Windows calculator again (in programmer mode). Make sure you have the *Dec* and *Byte* options activated (a byte is 8 binary digits (bits) and each binary digit is a 0 or a 1).

Q03.08: What is the largest positive number you can enter?

Q03.09: What is the largest negative number you can store? (Hint: try repeated subtraction from 0.)

Q03.10: What happens when you add 80 to 90? This is called *overflow* (when a number becomes too big)?

Q03.11: Give an example calculation that will cause *underflow* (when a number becomes too small).

Q03.12: Fill in Chart 3.2.

Chart 3.2 Answer form for Q03:12

	Storage size: Word (16 bits)	Storage size: DWord (32 bits)	Storage size: QWord (64 bits)
Maximum (positive) integer allowed			
Minimum (negative) integer allowed			

Abstraction 3: Integers to Floating Point Numbers (i.e. Rational Numbers)

Rational numbers, defined mathematically as one integer divided by a second integer, can be represented in three different ways (in base 10): a ratio of integers, a group of symbols with a decimal point (e.g., "7.543"), or in exponential notation (like scientific notation). If we represented rational numbers as a ratio of integers, requiring the computer to track the numerator and denominator would require very complex hardware to do the arithmetic and would be too slow. Using group of symbols require converting each number into a set of symbols from the 0's and 1's – a very slow process which effectively makes our computer no longer a binary machine. But if we use exponential notation, we can also adjust the exponent so the decimal point "." (radix point if working in bases other than 10) is always right before the first non-zero symbol and therefore it need not be stored.

We can now represent a floating point number with three categories (see Fig. 3.1): (A) the sign (positive or negative, (B) the mantissa (also called the fraction), and (C) the exponent.

We are now left with two limits for representing rational numbers: the number of digits in the mantissa (limits on accuracy) and the magnitude (positive exponent - > very large, negative exponent - > very small) of the number. IEEE has defined the number bits for each component dependent on how many bits the computer uses to represent the rational number, 32 bits or 64 bits, typically. For a 32 bit number, there are seven significant digits and the exponent can range from approximately −38 to +38. For 64 bits, we can have up to 16 significant digits and the exponent

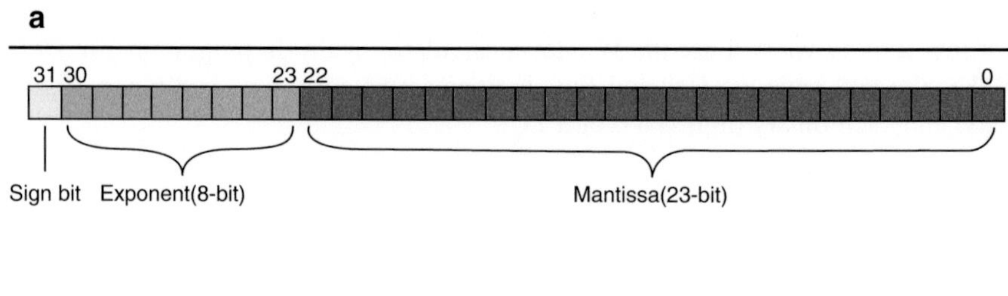

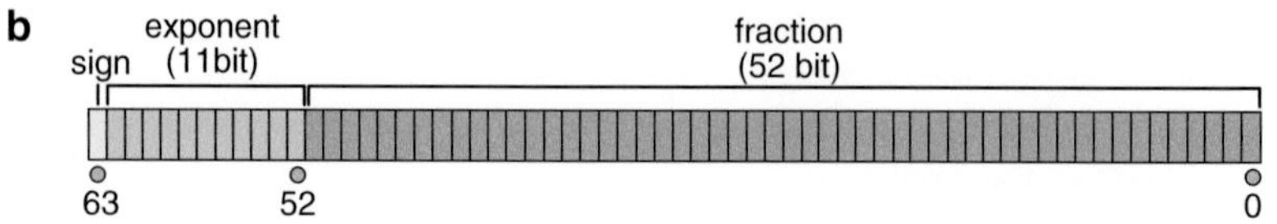

Fig. 3.1 Floating point bit representation. (**a**) 32-bit (**b**) 64-bit (taken from http://diaryofagraphicsprogrammer.blogspot.com and http://en.wikipedia.org/wiki/IEEE_754-1985)

can range from −308 to +308. While most machines, based on these standards, will do 32 and 64 bit arithmetic, there are also machines that use 128 bits (whose definition can be found on the internet). These machines will likely become more popular as time goes on.

Answer It!

Q03.13: Why are 23 bits equal to seven significant digits?

Q03.14: What is the largest number that can be represented with 32 bit floating point? The smallest?

3.6 Limits and Errors

There are a number of times when the computer's arithmetic does not yield the same results as algebraic arithmetic because of the finite nature of numbers stored in computers.

With the base ten number system (decimal), it is impossible to represent 1/3 as a floating point number accurately. It is a *repeating decimal*. We have a similar problem (only worse) when working with binary. Any number that cannot be evenly divided by a power of 2 (and there are more of these than there are in base ten) cannot be stored accurately in a computer, no matter how many bits are used. We also have irrational numbers (e.g. Pi and the square root of 2) that also cannot be represented accurately. Both of these types of numbers will introduce error into any representation of them on the computer, and in any subsequent calculation.

Discuss It!

List the different types of errors in addition to underflow and overflow (using examples) that you think can occur when working with floating point numbers.

Answer It!

Q03.15: Give an example of a very large number of items that you might run across in a field of science. Try to come up with the largest number possible. How big is it? What size of an integer (or floating point) is needed to represent it on the computer?

Q03.16: Fill in Chart 3.3 based on what is the smallest type of number needed to represent the item.

Chart 3.3 Answer Form for Q03.16

Item description	Number	Size (16 bit integer, 32 bit integer, 32 bit floating point, 64 bit floating point)
Number of people in the world		
Number of internet devices needed (assuming that no one person will ever have more than 100 devices)		
The age of the universe in seconds		
The number of atoms in a human body		

In order to study floating point numbers, you will use an Excel workbook (Module03.xlsx) and a Java program called Speed, both downloadable from the textbook website. One thing you need to know about Excel is that it stores all numbers internally as floating points. You can change only how it is displayed. You have no control over how it is stored (number of bits, size of the mantissa, etc.).

The first worksheet in the Module03.xlsx workbook, *Representation error*, deals with how a number is represented. It computes Y^X. Algebraically, the quantity (A-1-A) must always be −1. However, because of the limitations of a computer, you may not get that answer. Increase the number of rows (different powers for X) by copying the last row a number of times (a large number is good). Work with different values of Y (start with integers greater than 1) to find when this equation (A-1-A) does not yield a correct answer.

Answer It!

Q03.17: What values of X and Y did you use to get a wrong answer?

Q03.18: What values of X and Y (if Y is not an integer) give you a wrong answer?

Q03.19: Expand the size of column B and use 2 for Y. How many digits of accuracy do you have until you get a mistake? (Hint ->any power of 2 must end in 2, 4, 6, or 8.)

3.7 Order of Operation

We learned in algebra class that $A+B+C=A+C+B$. (What is the name of this mathematical property?), and you probably have seen various "challenges" posted on social media related to order of operations. Not only do we need to know what the order of operations is for the computing environment we use (so we don't end up with an unexpected answer), but because of our limits in precision, the order of the operations (which is not the same as the "order of operations"!) can be very important when working on a computer.

Answer It!
Use the Windows calculator for the first two questions that follow:

Q03.20: It is important to know when integer arithmetic is being done. What is the result of 3/2*5 while in DEC/BYTE mode? What is the result of 3 * 5/2 when done in scientific mode? Why are the answers different?

Q03.21: What are two possible arithmetic errors you have experienced with integers during this activity? Give a new example of each.

Q03.22: If X = 355/113, Y = 101/113, and Z = 52/113, what is X-Y-Y-Y-Z? What is X-Z-Y-Y-Y? What is X-2*Y-Y-Z? Enter these three formulas in the appropriate boxes in the *Error based on order* worksheet in the Module03.xlsx workbook. Report the results.

Q03.23: Based on your previous results, what can you conclude about the importance of the order of the operations? (Note: answers may be different depending on the use of a windows machine or mac machines and which spreadsheet software you use -> not all spreadsheet software use the hardware arithmetic for a number of different reasons).

Often a small error may seem to be so small that it does not matter. But errors can accumulate to the level that it does become significant.

Answer It!

Q03.24: The *Accumulating error* worksheet in the Module03.xlsx workbook computes 10-.2n two different ways. Column B computes 0.2 * n and then subtracts. Column D subtracts 0.2 for each row. The answers should be (and appear to be) the same. Find the first row with a mistake.

Q03.25: Simply subtracting the d row from the b row (example: =B2-D2) will not show the problem (unless you increase the number of digits displayed). Expand the number of digits shown for B and D -> which should be the differences. Which rows have mistakes?

Q03.26: Which column appears to be the most accurate? Which cells appear to be correct but might still have a mistake? (Hint: change the number of digits displayed.)

Q03.27: How many different types of errors can occur when working with floating point numbers? Give an example of each.

3.8 Accuracy and Speed

If you want to make your computer "more" accurate, you just need to store more digits. As a matter of fact, there are routines that programmers can use to make their arithmetic "perfect" when dealing with rational numbers. They use storing the number as a fraction and having as many digits as necessary. However, there is always a cost. Any increase in accuracy will usually cost in speed.

Program Speed available on the textbook website is a java program designed to demonstrate the difference in time that is required to do the same or similar operations on different data types. It compares the time required for three basic data types: integer, single precision floating point, and double precision floating numbers. It will also allow you to compare addition or multiplication speeds. Computer speeds will depend on other tasks the computer is doing. So while you run the program with different options, you will get the best results if you keep all other activities the same for each option. It also displays the answer. The program is best run from the command line (see your instructor for more information).

Answer It!

Q03.28: Run the Speed program and fill in Chart 3.4. Be sure to be very patient and don't press a button more than once! (You can work on other parts of this module while you are waiting for the calculations to finish.)

Q03.29: Which results are more accurate when using floating point numbers (either size)?

Q03.30: Which results are faster?

Chart 3.4 Answer Form for Q03.28

Data type and operation	Time required	Accuracy
Integer addition		
Integer multiplication		
Floating point addition		
Floating point multiplication		
Double precision addition		
Double precision multiplication		

Dec	Hx	Oct	Char		Dec	Hx	Oct	Html	Chr	Dec	Hx	Oct	Html	Chr	Dec	Hx	Oct	Html	Chr
0	0	000	NUL	(null)	32	20	040		Space	64	40	100	@	@	96	60	140	`	`
1	1	001	SOH	(start of heading)	33	21	041	!	!	65	41	101	A	A	97	61	141	a	a
2	2	002	STX	(start of text)	34	22	042	"	"	66	42	102	B	B	98	62	142	b	b
3	3	003	ETX	(end of text)	35	23	043	#	#	67	43	103	C	C	99	63	143	c	c
4	4	004	EOT	(end of transmission)	36	24	044	$	$	68	44	104	D	D	100	64	144	d	d
5	5	005	ENQ	(enquiry)	37	25	045	%	%	69	45	105	E	E	101	65	145	e	e
6	6	006	ACK	(acknowledge)	38	26	046	&	&	70	46	106	F	F	102	66	146	f	f
7	7	007	BEL	(bell)	39	27	047	'	'	71	47	107	G	G	103	67	147	g	g
8	8	010	BS	(backspace)	40	28	050	(	(	72	48	110	H	H	104	68	150	h	h
9	9	011	TAB	(horizontal tab)	41	29	051	)	)	73	49	111	I	I	105	69	151	i	i
10	A	012	LF	(NL line feed, new line)	42	2A	052	*	*	74	4A	112	J	J	106	6A	152	j	j
11	B	013	VT	(vertical tab)	43	2B	053	+	+	75	4B	113	K	K	107	6B	153	k	k
12	C	014	FF	(NP form feed, new page)	44	2C	054	,	,	76	4C	114	L	L	108	6C	154	l	l
13	D	015	CR	(carriage return)	45	2D	055	-	-	77	4D	115	M	M	109	6D	155	m	m
14	E	016	SO	(shift out)	46	2E	056	.	.	78	4E	116	N	N	110	6E	156	n	n
15	F	017	SI	(shift in)	47	2F	057	/	/	79	4F	117	O	O	111	6F	157	o	o
16	10	020	DLE	(data link escape)	48	30	060	0	0	80	50	120	P	P	112	70	160	p	p
17	11	021	DC1	(device control 1)	49	31	061	1	1	81	51	121	Q	Q	113	71	161	q	q
18	12	022	DC2	(device control 2)	50	32	062	2	2	82	52	122	R	R	114	72	162	r	r
19	13	023	DC3	(device control 3)	51	33	063	3	3	83	53	123	S	S	115	73	163	s	s
20	14	024	DC4	(device control 4)	52	34	064	4	4	84	54	124	T	T	116	74	164	t	t
21	15	025	NAK	(negative acknowledge)	53	35	065	5	5	85	55	125	U	U	117	75	165	u	u
22	16	026	SYN	(synchronous idle)	54	36	066	6	6	86	56	126	V	V	118	76	166	v	v
23	17	027	ETB	(end of trans. block)	55	37	067	7	7	87	57	127	W	W	119	77	167	w	w
24	18	030	CAN	(cancel)	56	38	070	8	8	88	58	130	X	X	120	78	170	x	x
25	19	031	EM	(end of medium)	57	39	071	9	9	89	59	131	Y	Y	121	79	171	y	y
26	1A	032	SUB	(substitute)	58	3A	072	:	:	90	5A	132	Z	Z	122	7A	172	z	z
27	1B	033	ESC	(escape)	59	3B	073	;	;	91	5B	133	[	[	123	7B	173	{	{
28	1C	034	FS	(file separator)	60	3C	074	<	<	92	5C	134	\	\	124	7C	174			|
29	1D	035	GS	(group separator)	61	3D	075	=	=	93	5D	135	]	]	125	7D	175	}	}
30	1E	036	RS	(record separator)	62	3E	076	>	>	94	5E	136	^	^	126	7E	176	~	~
31	1F	037	US	(unit separator)	63	3F	077	?	?	95	5F	137	_	_	127	7F	177		DEL

Source: www.LookupTables.com

Fig. 3.2 ASCII Lookup Table (www.lookuptables.com)

Abstraction 4: Numbers to Characters

In order to represent a single character (text, a digit (as a symbol), or other symbols) using a number, we use a simple table look-up. There are two different tables that are often used. The smaller and older table is ASCII (American Standard Code Information Interchange) as shown in Fig. 3.2. ASCII tables use 7 or 8 bits depending on the implementation. The UniCode table is a more modern (and now common) lookup table to convert numbers to characters. It uses 16 bits and is backward compatible with the ASCII table (if the first 8 leftmost bits are zero, then the table is entries are the same as the ASCII table).

Answer It!

Q03.31: How many different characters can be represented with 8 bits?

Q03.32: How many different keys are there on your computer keyboard? Multiply it by two (to allow for the use of the shift key).

Q03.33: How many spaces are left in the table?

Q03.34: Is 8 bits sufficient to handle all the different languages and alphabets in the world?

Q03.35: How many different characters can be represented if 16 *bits* (a computer "word") are used?

Q03.36: Fill in Chart 3.5 with the number of bytes required to represent each item. Feel free to use the Internet to find the answers. Chart the results on a logarithmic scale in Excel. To make life easier, you can approximate and use the preceding prefixes if you wish:

Chart 3.5 Answer Form for Q03.36

Item	Size (in number of bytes)
KJV version of the Bible in ASCII	
KJV version of the Bible in Unicode	
One byte for each particle in a mole	
A list of every star in the milky way (assuming each star's name takes 32 bytes)	
200 bytes for every nucleotide in the genetic code of your favorite organism	

Abstraction 5: Combining Data

Very rarely is one piece of data (a single number or character) enough to describe a fact or piece of information. You almost always need to store a lot of pieces of data to describe something. This data takes on one of two forms:

(a) Homogeneous data (a list of similar data) – This might be a list of temperatures taken every minute over a 12 h period. In this case it is just a list of 720 floating point numbers to be stored. However, most scientific information is not based on a single list of numbers. Homogeneous data is not sufficient.

(b) Heterogeneous data (a group of data of different data types) – This might be a data about a chemical reaction that would include the formula representing the reaction (a string), the energy required to start the reaction (a floating point number), the number of chemicals involved (an integer), etc. This data might have different integers, floating point numbers, and characters. These are often combined into what is called a *record*. You might then have a list of records about specific instances of that organism.

Discuss It!

What different types of records that might exist for different areas of science? Things that might be included are information about the collection of the data, the quality of the data, etc.

Answer It!

Q03.37: What are three different examples of data in an area of science that could be collections of homogeneous data? (You might need to use simplified examples here because most scientific data is heterogeneous.)

Q03.38: What are three different examples of data in an area of science that could be collections of heterogeneous data?

3.9 Collecting Groups of Similar Data

Collections of similar data can be organized based on the number of dimensions. One dimension (Fig. 3.3) can be visualized as a vector, two dimensions as a table (Fig. 3.4), three as a cube (Fig. 3.5), four as a row of cubes (Fig. 3.6), five as a table of cubes (Fig. 3.7), etc. If this method is used, the amount of space required is the amount of space required for each element multiplied by each dimension, regardless if there is data in a spot or not. If there is a lot of missing data, there can be a lot of wasted space.

Fig. 3.3 One dimension representation

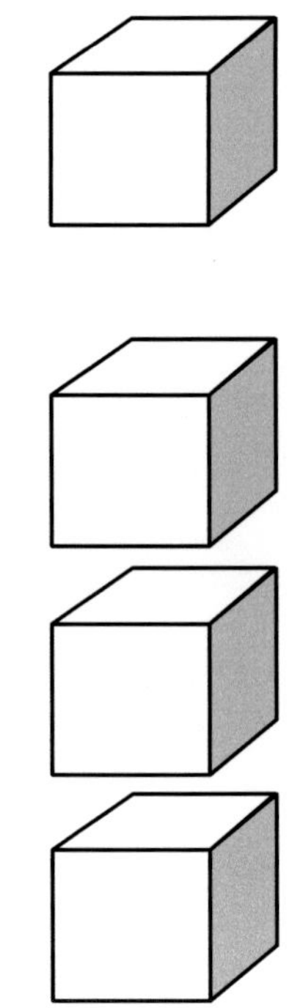

Fig. 3.4 Two dimension representation

Fig. 3.5 Three dimension representation

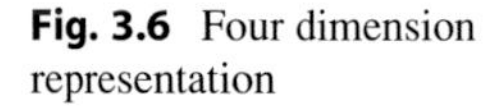

Fig. 3.6 Four dimension representation

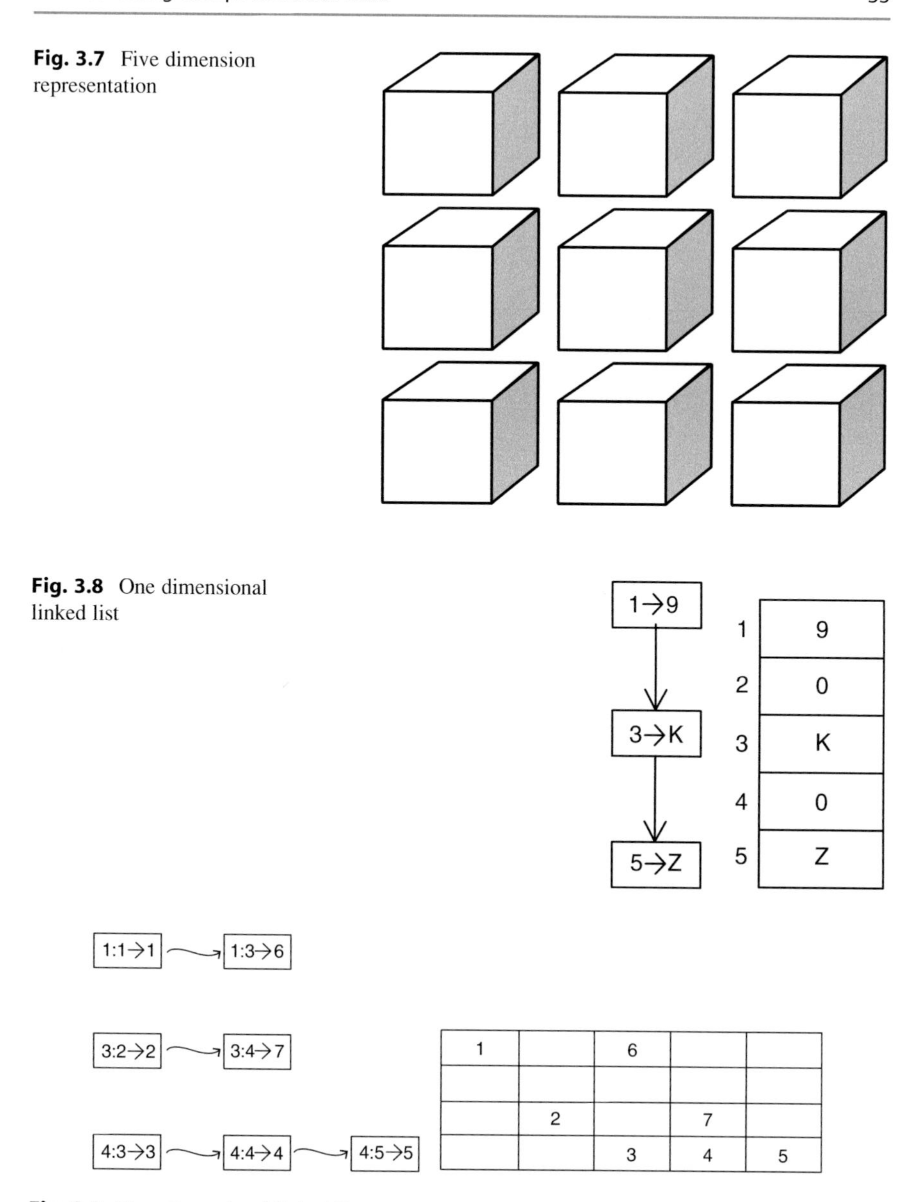

Fig. 3.7 Five dimension representation

Fig. 3.8 One dimensional linked list

Fig. 3.9 Two dimensional linked list

Another way to store a collection of similar data is a linked list. Each piece of data is stored in a *node*. Each node has the data and at least one pointer to the next piece of data. If the place in the list is important, then it will also store its *address*. A single dimension would look like a list of items with each item pointing to the next item (Fig. 3.8). Two dimensions would be a list of lists (Fig. 3.9), with three dimensions being a list of lists of lists (Fig. 3.10). Each node would require an additional 32 or 64 bytes (depending on the size of memory).

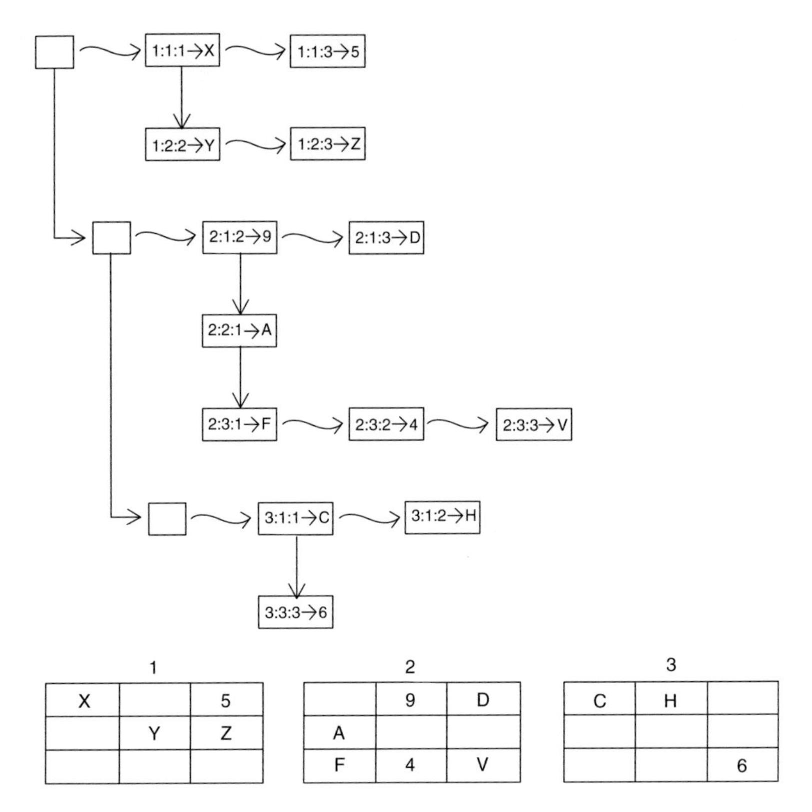

Fig. 3.10 Three dimensional linked list

Answer It!

Assume that you are modeling animals (birds and insects) found in the air (one cubic kilometer) at the resolution of 1 cm. Each cubic inch stores a 32 bit integer indicating which species, or part of a species, is in the cubic inch. You will be using each different bit to represent the presence or absence of one of 32 different species. 0 would indicate that the air is empty.

Q03.39: How much memory is required to store all the information as a 3 dimensional array?

Q03.40: What if you were using a three dimensional linked list instead? (Assume that it takes 32 bits to store the address of the next number.)

Q03.41: What if when using linked list, you decided not to store the 0's, and only 5 % of the air had something in it? (Don't forget you need to store the X, Y, and Z values – each 32 bits long.)

Q03.42: What percentage of the air needs something in it for both methods to require the same amount of memory? Create a graph in Excel to show the amount of memory needed for each method based on how dense the data is (i.e. the percentages of cells that are not 0).

3.10 Adding Structure to the Homogeneous: Trees

More meaning can be added to the data if a more complex structure is used. One such structure is called a tree. A tree is made of nodes and branches. Nodes have data and branches connect the nodes. There is one special node called the *root*. It only has branches going down to the children. It has no parents (branches going up). Each node can only have one parent. A general tree has no limit on the number of children a node can have. If a node does not have any children, it is called a leaf. Descendants are nodes that are children or descendants of children. Ancestors are a parent or an ancestor of a parent.

Answer It!

All the following questions relate to Fig. 3.11.

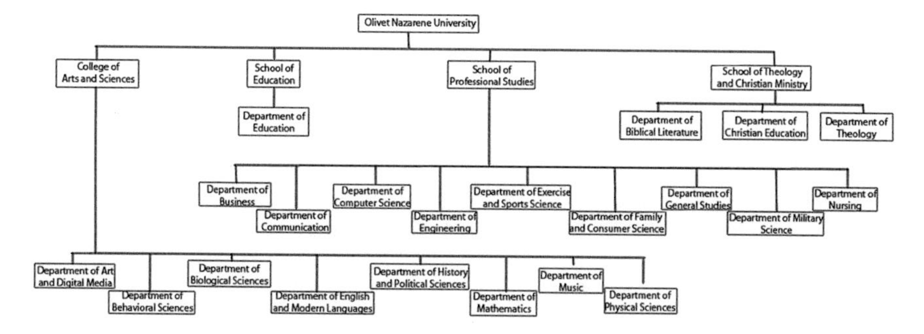

Fig. 3.11 Example Tree for University Academic Organization

Q03.43: What is the root?
Q03.44: What are all the leaves?
Q03.45: What are the ancestors of the Computer Science Department?
Q03.46: Which node has the most children?

One special type of a tree is called a *binary tree*. In this tree, each node can have at most 2 children. Because of this, you can refer to each child as a left or right child.

Answer It!

All the following questions relate to Fig. 3.12.

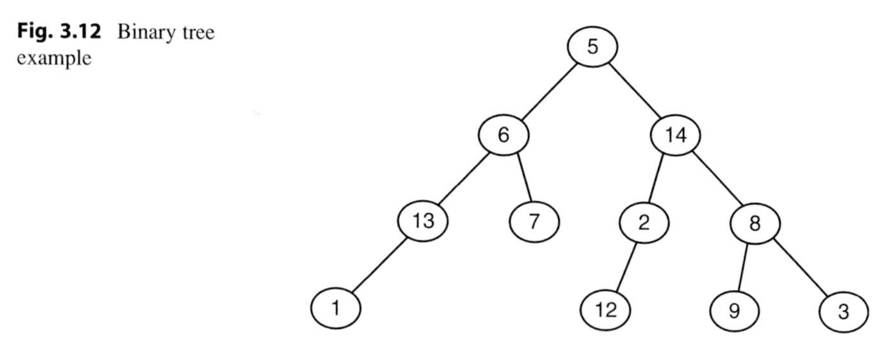

Fig. 3.12 Binary tree example

Q03.47: What is the root?
Q03.48: What are all the leaves?
Q03.49: What are the ancestors of 1?
Q03.50: Which are the descendants of 14?
Q03.51: What is the right child of 8?, 13?
Q03.52: What is the left child of 2?

3.11 Adding Structure to the Homogeneous Collections: Graphs

Trees are great data structures for representing information with a hierarchical structure, but not all knowledge has such a structure. Another structure that is common is a graph. By removing the restriction of only one parent per node for a tree, we now have a graph. When this is done, there is no longer the idea of parents and children. One example of a graph (Fig. 3.13) might be the cities that are connected for a small airline. Each node would be the city and each edge would represent a flight between the two cities. If the edge has a weight (such as the miles), the graph would be called a *weighted graph*. The edges may also be unidirectional (only go one way) as seen in Fig. 3.14. This is called a directed graph (or digraph). Formally, Fig. 3.14 would be called a *directed weighted graph*.

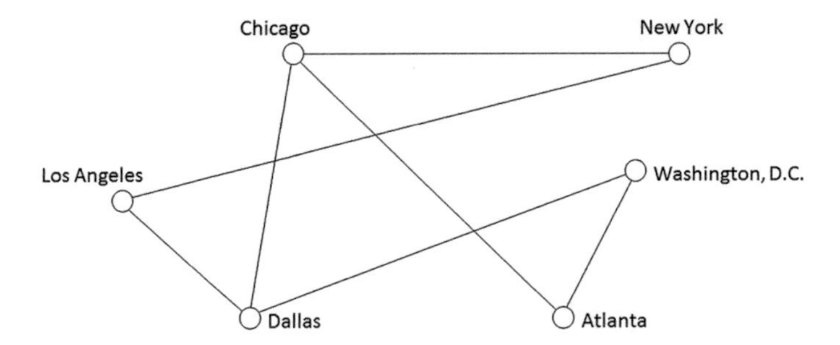

Fig. 3.13 Graph example

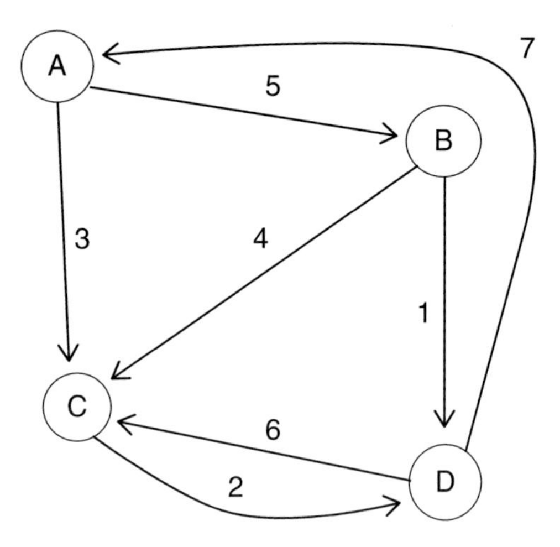

Fig. 3.14 Directed weighted graph example

Both graphs and trees can be represented in the computer. They are usually done with linked lists (but can be represented with an array or vector, respectively). But because of the use of abstraction, most of the time we don't worry about how they are represented.

3.12 Adding Structure to Homogeneous Collections: Stacks and Queues

Two other data structures that are easy to understand are stacks and queues. You use them all the time in real life. One very common example of a stack is a stack of dishes. The rule (unless you are a magician) is that you only place dishes on the top of the stack (called *push*) and you only take them off the top of the stack (called *pop*). Because of these two roles, stacks have a unique characteristic: *Last In, First Out* (LIFO). The following commands are modeled in the Fig. 3.15.

Create a stack, Push *5*, Push *4*, Push *3*, Pop, Pop, Push *2*, Pop, Pop.

Answer It!

Q03.53: Draw a similar diagram for the following instructions:

Q03.54: Create a stack; Push *Black*, Push *Green*, Pop, Push *Red*, Push *Green*, Pop

Discuss It!

What type of things could go wrong with a stack?

Queues are similar to stacks except that you take items off the *head* of the list (dequeue) and place them on the end of the list (enqueue). This characteristic is called *First In, First Out* (FIFO). Hopefully the line for getting into the dining room acts like a queue! The following commands are modeled in the Fig. 3.16.

Answer It!

Draw a similar diagram like that shown in Fig. 3.16 for the following instructions: Create a queue, Enqueue *Black*, Enqueue *Green*, Dequeue, Enqueue *Red*, Enqueue *Green*, Dequeue

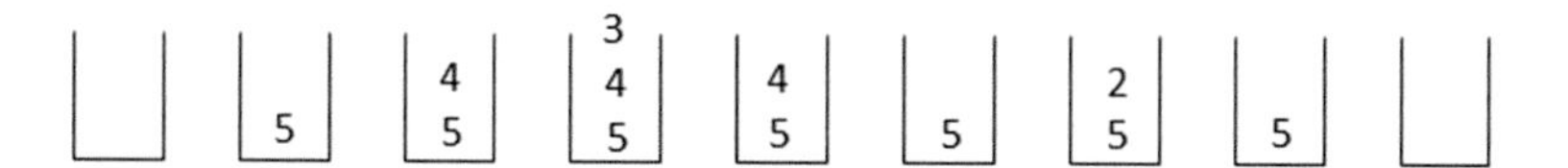

Fig. 3.15 Queue example

Fig. 3.16 Visualization of the commands: *Create queue, Enqueue 5, Enqueue 4, Enqueue 3, Dequeue, Dequeue, Enqueue 2, Dequeue, Dequeue*

Q03.55: How does this look/act different from the stack activity?

Discuss It!
What type of things could go wrong with a queue?

Answer It!

Q03.56: What are three examples of trees in an area of science? Which (if any) are binary trees? (If you need to, you may use other areas of science.)

Q03.57: What are three examples of graphs in an area of science? Which (if any) are directed graphs? Which (if any) are weighted graphs? Which (if any) are weighted directed graphs? (If you need to, you may use other areas of science.)

Q03.58: What is an example of a stack in an area of science? (If you cannot think of one, try another area of science.)

Q03.59: What is an example of a queue in an area of science? (If you cannot think of one, try another area of science.)

3.13 Related Modules

- Module 1: Introduction to Computational Science. Computational science and errors are covered.
- Module 4: Scientific Data Acquisitions. Errors, space, and precision issues are evaluated.
- Module 6: Solving Equations. Errors are discussed.
- Module 10: Self-Defining Data: Compression, XML, and Databases. Space issues are explored.
- Module 12: Curve Fitting. Errors are discussed.

References

Neeley M, Ansmann M, Bialczak RC, Hofheinz M, Lucero E, O'Connell AD, Sank D et al (2009) Emulation of a quantum spin with a superconducting phase Qudit. Science 325 (5941):722–725. doi:10.1126/science.1173440

Sample, Ian, and Science Correspondent (2007) Einstein + Bacteria DNA = Organic Computer Breakthrough. The Guardian, February 22, sec. Science. https://www.theguardian.com/science/2007/feb/22/uknews.sciencenews

"Scientists Create Organic 'Molecular Computer'" (2016) Accessed 27 June. http://www.gizmag.com/organic-molecular-computer/15041/

Scientific Data Acquisition

4

4.1 Objectives

After completing this module, a student should be able to:

- Define the terms calibration, precision, accuracy, instrument drift, and resolution.
- Appreciate the difficulty of scientific measurements using instrumentation.
- Construct and execute an experimental procedure that uses instrumentation.

4.2 List of Terms

- Accuracy
- Calibration
- Drift
- Precision
- Resolution
- Sampling frequency
- Sampling interval
- Sampling length

4.3 Motivation

Science and engineering relies on data. Scientists and engineers analyze and interpret data using many of the tools and techniques covered in this course. But every scientist and engineer needs to remember that all data is collected with instruments – from something as simple as a ruler, to something as complex as a linear accelerator. No matter what instrument is used, there are limits to the resulting data (e.g. precision, accuracy, instrument range, sampling frequency,

K. Brewer, C. Bareiss, *Concise Guide to Computing Foundations*,
DOI 10.1007/978-3-319-29954-9_4

etc.). Those limits will affect computational results and subsequent analysis/interpretation. In this module, you will be exploring these issues.

4.4 A First Problem – Introduction

This module assumes you have access to a Vernier Instrument's LabQuest unit with appropriate sensors. Refer to the Appendix for an overview of the Vernier LabQuest instrumentation equipment and general procedures. You may be able to do similar activities with other devices. Please refer to your instructor about which device to use.

Assume you want to collect data to answer the question: "What is the temperature of the classroom?" This is a seemingly simple question...

Discuss It!
What are some of the issues that you may need to consider when collecting data to answer the above question?

Answer It!
You will be using a LabQuest unit and a Stainless Steel Temperature probe to collect room temperature data. *Do not touch the metal part of the temperature probe sensor*. Plug the sensor into the LabQuest unit and turn on the device.

Q04.01: Soon after unit powers up, record the temperature reading. Be sure to record your answer with the given number of significant figures.

Discuss It!
You may notice that the measured value is not constant. What do you think is causing the change? How might you account for that fluctuation in reporting significant figures? What might you do to increase the precision of your measurement (without changing instruments!).

Q04.02: Now press the *collect* button on the LabQuest. Your unit will default to 180 s of data and will graph it for you as it is collected. Use the built-in statistical analysis process to determine the mean temperature value and the standard deviation. Record both values.

Q04.03: Based on your answer of the average temperature and standard deviation, how many significant figures should you really report for the average temperature?

Q04.04: Press the *collect* button on the LabQuest again, and this time walk around the room with your instrument. Your unit will default to 180 s of data and will graph it for you as it is collected. Export your data and plot it in Excel and report your plot.

Q04.05: What was the maximum, minimum, and average temperature recorded?

Q04.06: If there are other students performing this data collection at the same time, compare your results to the other's results. Comment on the observed differences and similarities. (If there are no other students, repeat your measures and analysis a few times using different walking paths. Your device may ask what to do about the previous data – feel free to discard it.)

Discuss It!

The precision of your previous temperature reports were reflected by the number of significant digits you reported. How did that previous precision compare to the variability (between different students, or your multiple measurements) of the measured values?

You should be able to now conclude that when reporting instrument measurement values, the precision of the instrument is not necessarily the precision of the measured value. You often need to report the statistical information from your measurements, not just a single number with a given number of significant figures.

Answer It!

Q04.07: Given all the measurements and analysis above, what now would be your answer to the question "What is the temperature of the classroom?"

4.5 Sensor Considerations

We will now consider the reaction of the sensor to the environmental condition. Sensors use physical and/or electronic implementations of scientific principles in various ways to determine the environmental property being measured (like temperature). Each technology will have differing costs, reaction speeds, accuracy, precision, and viable ranges – no matter the value being measured. Scientists and engineers need to consider each when selecting a sensor during their data collection activities. Four common technologies to measure temperature are with thermistors, thermometers, infrared, or thermocouples.

Answer It!

Q04.08: Back at your seat, prepare to collect another 180 s temperature dataset. Start recording data and after about 30 s, grab the end of the metal temperature probe with your bare hand. Hold it for about 10 s, then let go of the temperature probe. Finish collecting the dataset. Export your data and plot it in Excel and report your plot.

Q04.09: Did the sensor stabilize at a higher temperature by the end of the 10 s of you holding the end of the probe? Assuming an average body temperature of 98.6 °F (37 °C) what was the difference between the sensor value and your body temperature? How long do you think it would take for the sensor to accurately measure your body temperature?

Q04.10: How long did the sensor take to "stabilize" to the room temperature after you let go of the probe?

Discuss It!

Lookup the basics of thermistors, thermometers, thermocouples, and infrared technologies to measure temperature. Consider expected temperatures, required precision, and reaction times, which technology would be more appropriate to measure the temperature of a gas flame for a hot water heater? Air in a room to control air conditioning/heating? The surface of a space capsule during reentry? A turkey being cooked in an oven?

4.6 Computing Issues

Commonly, sensor data is collected and process with computing technology. As covered in Module 03, values can be stored in a variety of formats/types (int, float, double, etc.) and in a variety of ways (lists, stacks, databases, etc.). Scientists need to understand and consider these issues when collecting and processing sensor data.

Answer It!

Q04.11: Using your knowledge of the temperature sensor and LabQuest accuracy so far, what data type do you think is used to store temperatures? Explain your answer.

Q04.12: Given your previous answer, what do you think should be the maximum temperature allowed? Minimum temperature?

Q04.13: Compare your previous answer with the maximum and minimum temperature according to the sensor data sheet. Is there a better data type that could be used to store the data considering the maximum and minimum temperatures on the data sheet?

Q04.14: Fill in the Chart 4.1. Use published values from the manufacturer when appropriate and available. Remember to include appropriate units.

Chart 4.1 Answer Form for Q04.14

	Temperature	Motion	Force
Minimum measured value			
Maximum measured value			
Resolution			
Accuracy			
Response time			
Max. value allowed			
Min. value allowed			
Accuracy attainable			
Suspected data type			

4.7 A Second Problem – Design

Let us now consider the problem of wanting to collect data to answer the question: "What is the temperature of the cold water and of the hot water available in this building?" At first glance, it may appear that this is another seemingly simple question. . .

Discuss It!

What might make the measured temperature (reported instrument value) change over time? What might make the actual temperature of the fluid change over time? How might you design your data collection procedure to identify and measure these changes? How long might you need to measure until your instrument gives you a correct value? What should be the physical setup so you can measure the temperature of the water flowing out of the faucet?

Answer It!

Q04.15: Write up a simple procedure (your methodology) for measuring the hot and cold water temperature in your building. Be sure to include a description of the physical setup for your sensor, the sampling frequency, the sampling length (be sure to take into account data storage limits of your LabQuest instrument/sensor), and your post processing (graphing and statistical analysis) to determine what information to report (your answer to the question).

Q04.16: Based on your documented procedure from the previous question, measure the temperature of cold water and hot water coming out of a faucet. Write up a short report of your experiment. Include all the data you collected, any deviations from your procedure, and your reported results that answer the original question ("What is the temperature of the cold water and hot water available in your building?").

4.8 A Third Problem – Bonus

[This section could be assigned as part of a term project.]

Answer It!

Q04.17: Create and perform your own experiment, subject to the following requirements:

- Use a minimum of two sensors (can be same type).
- Measure something in the real environment (i.e., not from an experimental/controlled setup).
- Measure for longer than 3 h.
- Generate a graph and analyze it.

Write a report summarizing your experimental question, methodology, data collected, analysis, and conclusions.

4.9 Computing Questions

- For the different integers in your module, what data size(s) do you think were used to represent them? Why?
- For the different floating point numbers in your module, what data size(s) do you think were used to represent them? Why?
- What are the different sources of errors that might be encountered in the computational work done in your module?
- For the largest typical size of data used in the area of science studied in your module, how much space is needed? Why?

4.10 Related Modules

- Module 3: Data Types: Representation, Abstraction, Limitations. Computational limits/errors (e.g. round-off, overflow, underflow, limited precision, non-reproducible computations) are examined further.
- Module 13: Curve Fitting. Algorithms are explored and explained.

5 Procedures: Algorithms and Abstraction

5.1 Objectives

After completing this module, a student should be able to:

- Read and understand simple NetLogo models.
- Make changes to NetLogo procedures and predict the effect on the simulation.
- Name the control structures sufficient to express all components of programs.

5.2 Definitions

- Algorithm
- Procedural abstraction
- Procedure

5.3 Motivation

Part of calling science and engineering "disciplines" is the implication that they often follow structured processes and procedures for data collection, analysis, and evaluation of their respective topic areas. Thus, scientists and engineers often create and/or follow those processes and procedures, many times using computing. But before learning more about procedures and computers and their abstraction, we need to first take a brief look at the concept of an algorithm. An *algorithm* is a mathematical term for a clear set of instructions that, when followed, solves a particular problem. Examples include methods you may have learned in school and life:

- Multiplying large numbers by hand
- Constructing a perpendicular to a line using a ruler and compass

K. Brewer, C. Bareiss, *Concise Guide to Computing Foundations*,
DOI 10.1007/978-3-319-29954-9_5

- Constructing an angle bisector using a compass
- Brushing your teeth
- Preparing a hotdog in a microwave

Discuss It!
Consider how you multiply two large numbers by longhand. Could you teach someone this algorithm?

To get the correct result from an algorithm, you must understand each instruction and carry them out in the right order. You also need some basic knowledge such as a *times table* (for multiplication) or entering cooking time (on a microwave). In these examples, a human is performing the algorithm in order to solve a problem. The key to an algorithm (i.e. set of instructions) is that it can work with many different inputs (data values) and therefore be reused and repeated. Algorithms are used in science and engineering disciplines to solve common mathematical problems.

5.4 Procedures

Programmers sometimes use existing algorithms or they may design algorithms of their own to find solutions. Once a clear set of instructions to solve a problem exists, they must be written in a programming language so the algorithm can run on a computer. The entire algorithm written in a programming language is called a program or set of *procedures*. The computer must "know" some primitive things like numbers and simple math operations, as well as various *procedural* instructions.

A program can be separated into parts called subprograms. The *structured program theorem* (Böhm and Jacopini 2011) states that every algorithm can be implemented in a programming language that combines subprograms in only three specific ways, called "control structures". These three control structures are:

- Executing one subprogram, and then another subprogram (*sequence*).
- Executing one of two subprograms according to the value of a Boolean condition (*selection*).
- Executing a subprogram repeatedly while a Boolean condition is true (*repetition*).

We will learn about algorithms, procedures, and programming by using the NetLogo (Wilensky 1999) system and programming language. The objective is not for you to become a good NetLogo programmer, but to have some experience and practical examples of a programming language using a defined set of built-in commands and control structures.

5.5 Control Structure Example

Say we want to create a repetitive spiral graphic that looks something like that shown in Fig. 5.1.

We can create this using NetLogo using a single turtle (the NetLogo term for an agent that can be programmed to move around the simulated world) with the pen down. All three control structures in the structured program theorem are needed, and the entire algorithm can be written in a single command:

repeat 72 [ifelse heading mod 10 < 5 [set color yellow] [set color green] repeat 4 [forward 10 right 90] right 5]

Answer It!

Q05.01: What part(s) of the command (algorithm) reflect a *sequence*?
Q05.02: What part(s) of the command (algorithm) reflect a *selection*?
Q05.03: What part(s) of the command (algorithm) reflect a *repetition*?

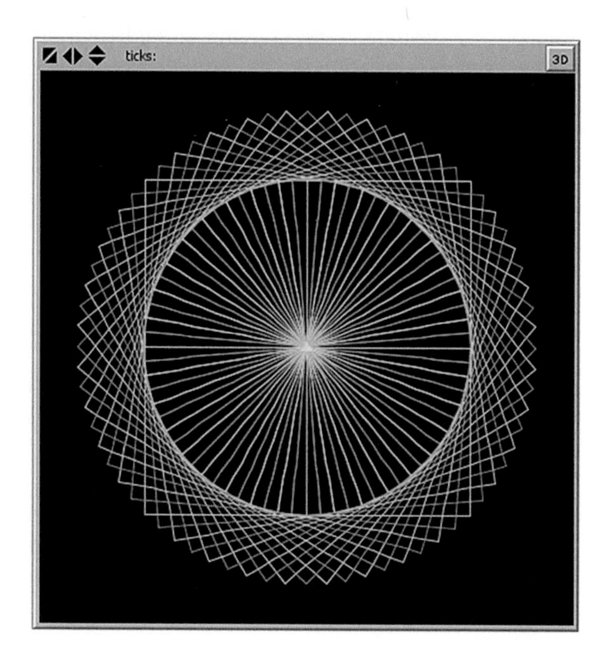

Fig. 5.1 Spiral graphic example

5.6 Procedural Abstraction

By now you have used examples of NetLogo control structures for each of the three types: *sequence, selection,* and *repetition*. There is one more powerful programming concept to cover – *procedural abstraction*. Because it is difficult to figure out exactly how to write a complex procedure and you often want to use them more than once, most programming languages have a way to give a procedure a name and to invoke it when needed. Once a name is associated with the procedure, you can run all the instructions in the procedure just by using the associated name. It's like a new command and can be used just like built-in commands. By giving a procedure a name, the set of commands is larger and the programming language is more expressive and more abstract.

Answer It!

Q05.04: Write a NetLogo procedure to have a turtle draw an equilateral triangle with sides of length 10.

5.7 Theater Lights Part 1

This model was inspired one day while driving past a local movie theater and watching the flashing lights moving around each sign. You will create the user interface and will program the algorithms that will perform the model behaviors. It is strongly suggested you read through the NetLogo Tutorial appendix prior to continuing in this module.

We will approach creating the model in two major efforts:

- we will first arrange turtles (our lights) around the edges of a 20×12 rectangle on every other patch.
- we will then make them move counter-clockwise to simulate the moving flashing.

Let us begin. Start NetLogo and create the basic *setup* and *go* buttons as instructed in the appendix. Ensure the *setup* procedure includes the "clear-all" and "create-turtles 1" commands.

The next thing to do in the setup procedure is to ask the only turtle to set its (x, y) location to (10, 6) the upper right corner of the rectangle. At the same time, we should set the color to *white* and make its heading north.

```
ask turtle 0 [
 set xcor 10
 set ycor 6
```

```
    set color white
    set heading 0
  ]
```

Try the code in the *setup* button and make sure it does what you've planned.

A turtle can make a copy of itself by using the command *hatch 1*. You'll use this command with the first turtle to leave lights along the path as it travels around a rectangle in a counter-clockwise direction. Remember that you plan to put a light on every other patch along the outside of a 20×12 rectangle.

To add the top row of lights, try adding these commands inside the *ask turtle 0* sequence (after the previous code).

```
  left 90
  repeat 10 [
   forward 2
   hatch 1
  ]
```

Test your modified procedure. Make sure you get it working (the top row lights are created) before you continue. Try sliding the speed slider left to slow the simulation down. Test your procedure again and watch it doing each command one at a time in response to pushing the *setup* button.

You should save your model periodically while you are modifying it. Do that now by selecting *File* and *Save* and naming your model (e.g. Lights1.nlogo).

Now you need to add commands to the *setup* procedure to complete the rectangle. The rectangle is 20×12. You repeated 10 times along the top because you want a light every other patch. You should therefore repeat 6 times on the sides. Each turn at the corners is 90° to the left. Don't continue until your *setup* procedure places all the lights around the rectangle. The corners of the rectangle should be at (10, 6), (−10,6), (−10, −6), and (10, −6).

You should add one more command to the end of the *setup* procedure. The turtle that traveled around the rectangle leaving copies of itself along the way using the *hatch 1* command will end up at its starting location in the upper right corner. But there is also a copy of it in that same starting location. To avoid duplication, we need to remove the turtle that hatched all the others. To remove turtles from the world, we use the command *die* to ask a turtle to leave this world. Therefore ask the turtle to "die" at the end of the "ask turtle 0" command. That should be all for the *setup* procedure.

Answer It!

Q05.05: Report your setup procedure to create the initial lights around the rectangle.

Now you need a way to have the lights change color and move around to the left (i.e. counter-clockwise) and you will create those commands in the *go* procedure. Remember that the procedure starts with "to go" and ends with "end".

Let us first modify the *go* procedure to blink the lights from white to yellow and back again. It possible to identify a set of agents (i.e., all the turtles) using the term "turtles". You previously gave commands to a single turtle agent with a command like ***ask turtle 0 [set color white]***. You can give commands to each turtle in the set of all turtles with a command like ***ask turtles [set color yellow]***. Add commands to set all the lights (turtles) to white and then a separate command to set all the lights to yellow.

Test *go* by selecting the *Interface* tab and pushing the *go* button several times. Try slowing the speed slider way down again until you can see what is happening to each light when you push the button. You should see that the turtles are in a random order when they are given commands by the ***ask turtles []*** command. It is a form of repetition control structure. Return the speed slider to normal now.

Now let us further modify the code to move the lights. Add a command to one of your ***ask turtles []*** commands to move your lights forward one distance. Test your code by running it.

Your model should be getting closer to what you want, but the lights are forgetting to turn left at the corners. You need a selection statement to see if a turtle is at one of the corners. If it is, have it make a 90° left turn. Then have it move forward 1. The way to see if a turtle is at the upper right corner (10, 6) would look like this:

```
if xcor = 10 and ycor = 6 [left 90]
```

Add a statement like this for each of the other three corners of the rectangle. Test your changes and see how it looks. Make sure you have also added the command ***tick*** to the end of the *go* procedure. This adds one to the ticks counter on the model to tell how many times the *go* procedure has run. Remember you can start the model over anytime by pushing the *setup* button. Adjust the speed slider until it looks like flickering theater lights. Make sure to save your model.

Answer It!

Q05.06: Give an example of a *sequence* of commands from your theater lights model.

Q05.07: Give an example of a *selection* control structure from your theater lights model.

Q05.08: Give an example of a *repetition* control structure from your theater lights model.

Discuss It!

Your model looks somewhat like flashing theater lights but you know that lights don't really move around the outside of a theater sign. How does a real sign achieve the look of moving lights?

5.8 Theater Lights Part 2

One way to make lights on signs appear to be moving is by turning them off and on in a sequence. We will change the previous model to model this technique. Open your previous theater lights model and use *File* and *Save As...* to make a copy of the model as Lights2.nlogo.

We will first modify the *setup* procedure to leave a turtle on every patch around the border instead of every other one. Make sure the lights create the same size rectangle. Test it and make sure it is working correctly before continuing.

After all the lights are on around the border, we now need to turn off every other light. Turtles have a Boolean condition variable ***hidden?*** which is either ***true*** or ***false***. Boolean variables end with a ***?*** to show that we are asking the question whether the turtle is hidden or not. Right click one of the lights and select the option to inspect whichever turtle you happened to click on. This should bring up the turtle monitor. Look to see that the ***hidden?*** variable has the value ***false***. Change it to ***true***. Did the light disappear? Change the value back to *false* and the light should come back on.

Remember that each turtle has a unique ***who*** number assigned when it is created (kind of like a social security number). Since the turtles are numbered in order during creation, the turtles with even *who* numbers will be every other one.

Even numbers are evenly divisible by 2. That means they have a 0 remainder when divided by 2. For each light that would be the condition ***who mod 2 = 0***. Now add commands to the end of the *setup* procedure to ask all turtles with an even who number to *set hidden? true*. (You may need to search the excellent NetLogo Dictionary online to determine the proper terminology to use.) Test your *setup* procedure until you get it working correctly (i.e. hides the even numbered turtles).

Answer It!

Q05.09: What was the command you used to hide the even numbered lights?

In order to give the illusion of moving lights, you want to keep turning lights on and off. For each light, if it is on, then turn it off. If it is off, then turn it on. One approach would be to use a statement like ***ifelse hidden? [set hidden? false] [set hidden? true]***. Study this code until it is clear.

Another method takes advantage of Boolean algebra operation ***not hidden?*** which gives the value *true* if ***hidden?*** is *false* and gives the value *false* if ***hidden?*** is *true*. This makes it possible to set the value of ***hidden?*** to its opposite. That would look something like ***set hidden? not hidden?***. Now modify the *go* procedure by removing the commands to move the turtles around the rectangle and replace with commands to turn the lights off and on each time *go* is run. Test your model until it works. Don't forget to save your model.

Answer It!

Q05.10: Does it look like the lights are moving around or just flashing on and off?

Q05.11: If the lights appear to be moving, which direction are they moving (clockwise or counter-clockwise)?

Q05.12: How might you improve the model to make it more realistic?

Discuss It!

Which theater lights model version do you think is better? Why? Which on was easier to implement? How much more effort and code do you think would be needed to make it more realistic?

5.9 Leaves on the River Part 1

This NetLogo model was inspired one beautiful day at the park from watching the river flowing by. Assume that it is fall and leaves are falling into the river on the left from where you sit watching. The river is flowing from left to right and you are viewing from above. All leaves in our section of the river will float down river (i.e. to the right) from where they enter on the left until they exit on the right.

Start NetLogo if it isn't already started and be sure to begin a *New* model. Begin by adding the basic, simple commands (see the appendix) and save the model with a good name (e.g. Leaves1.nlogo). Don't forget to periodically save your model as you modify it.

You have seen the expressive power of *procedural abstraction* in enabling us to name a set of commands. As you learn in Module 3, *data abstraction* provides power in naming data types. In the previous theater lights models, you represented lights with turtle agents. You may have found it difficult or at best annoying to always remember to call a light a turtle. Wouldn't it be nice to be able to call a light a *light*?

NetLogo provides data abstraction by having every turtle have variable *breed*. You can use the naming power of *abstraction* to create a new breed of turtle. If you add the statement: ***breed [leaves leaf]*** at the top of your code, this defines a new breed. Now you have a set of agents with the name ***leaves***, with individuals called

leaf. This one brief definition now makes it possible to use a command ***create-leaves 10*** instead of ***create-turtles 10***. You can also give commands to this new breed of turtles with ***ask leaves []*** instead of ***ask turtles []***. It merely makes it easier to read *leaves* and think *leaves* as well as think *leaves* and write *leaves*.

Lets uses this abstraction by creating leafs in the new model. The *create-leaves* command expects a number which will be the number of agents to be created. It can also accept a block of commands to give to each *leaf* as it is created. This is the same as creating some leaves and then asking them to all run a block of commands, but a little more convenient. Add these commands to your *setup* procedure:

```
create-leaves 10 [
  set color green
]
```

Now add commands to the *go* procedure to have all the *leaves* move with each *tick* of the simulation. Try running the model.

Abstraction has allowed the naming of procedures and data according to what they do and represent. You have seen the expressiveness of calling a procedure *draw-square* when that is what it does. Also, you can now call a leaf a *leaf*. Wouldn't it be better if in addition to calling a leaf a *leaf*, you could look at a leaf and have it look like a leaf instead of an arrowhead?

Select the *Tools* menu, then select *Turtle Shapes Editor*. There is a library of shapes available for turtles and you can create your own. Scroll down until you see the *leaf* shape. It is already included in the model. Select the *leaf* shape and push the *edit* button. Check the *rotatable* box near the bottom left and push the *OK* button to save the change to the shape. The arrowhead shape is the default shape of turtles. Close the *Turtle Shapes Editor*. Add a command to the *setup* procedure to ***set shape "leaf"*** for each of the *leaves* created. Doesn't that look better? It might be a little confusing with the word ***leaf*** meaning a type of turtle agent and ***"leaf"*** meaning a shape. Remember that the naming of the agent breeds can be anything and you choose them because they mean something to you.

Now when running the model, you should see all your leaves start at the middle and then move out in random directions. They continue to move in straight lines, and when they move past the edges of the world (the black window), they appear on the other side (top to bottom, right to left, etc.). This is not realistic for a river and we need to make some modifications to the placement and movement of our leaves.

Many real world phenomena occur with a particular probability. When observed, these phenomena appear to occur by random chance. Probability and statistics is the mathematical study of these types of occurrences. Mathematical models that represent random outcomes are often used to build what are called *Monte Carlo Simulations*. Much study of games with random chance, such as cards and dice, have contributed to this type of simulation model. You will use some commands to add realistic *randomness* to your model.

As leaves fall, they are usually a random set of colors. Try setting the color of the leaves with ***set color one-of [green yellow brown]***. ***one-of*** randomly selects one of the values in a list enclosed in brackets. Test it to make sure it works.

By default, when the leaves are created they are placed in the center of the world and have a random heading from 0 to 360. When they go *forward,* they all go in the random direction they are headed. You want them to be randomly placed in the world but all flow "downstream" to the right. We will first place the leaves randomly in the world, then we will make sure they all move to the right.

The leaves are currently all being placed in the center of the world at the origin. We need to set the xcor and ycor of our created leaves to a random coordinate within the world. The commands ***random-pxcor*** and ***random-pycor*** provide a random x and y value. Specifically ***random-pxcor*** returns a uniformly distributed random number in the range of all the patches x coordinates on the world grid, same for the other command for y coordinates. Test your code.

Moving to the right can be accomplished by two different ways: changing the heading of all the leaves to 90° and moving them forward 1 (or more) distance; or by changing all leaves *xcor* variable to be 1 more (which is the patch to the right). Let us use the latter and modify the *go* procedure to ask the leaves to ***set xcor xcor + 1***. Test your code.

Answer It!

Q05.13: Does your model look like leaves moving down a river? List things that are unrealistic and should be improved.

Q05.14: Report your model code.

The model currently just causes the leaves that move off the right side of the world to reappear on the left (at the same vertical location). This is somewhat unrealistic. We could improve the model by randomly changing the vertical placement of the leaf when it wraps to the left side. We would need to use a conditional structure along with a reset of the y coordinate. Note that you need to test if you are going to go off the right side before you move the leaf, as the system will immediately place the leaf on the right side.

Answer It!

Q05.15: Implement code to cause the leaves to reappear on the right side in a different vertical location. Report your model code. (Be sure to save your model.)

Discuss It!

An alternate way to improve the look of the model by having leafs appear on the right differently than where they leave. This can be accomplished by turning off the world wrapping, killing the leaving leafs, and creating a new leaf (even with a new color) on the left side. Try modifying your model accordingly.

5.10 Leaves on the River Part 2

Most computational science models have one or more input variables which are changed within some set of possible parameters. These are the independent variables for the experiment. Your current leaves on the river model currently has 10 leaves on the river. In order to vary the number of leaves, we must add an input control.

Select the *Interface* tab and then choose a *slider* from the drop-down list of controls. Add the slider control to your model interface above the *setup* button. A slider dialog should pop-up. Complete the dialog: Global variable: ***leaves-in***, Minimum: ***1***, Increment: ***1***, Maximum: ***30***, initial Value: ***15***, and Units: ***leaves***. Then push the *OK* button to save the slider control. This has defined a new global variable names ***leaves-in*** we can refer to in our code. You can change the slider input parameters anytime by right clicking on the slider and selecting *Edit*.

Next, we need to connect the slider control to the procedures code. Change the leaf creation command by replacing the ***create-leaves 10*** with ***create-leaves leaves-in***. This now creates the number of leaves as indicated on our slider at the start of the model. Run the model with different initial leaves. Change the slider to allow 100 leaves in the river.

Sometimes we want to monitor and record information from our model/simulation as it is running, often with a graph. Let us add a graph to our NetLogo model and have it report the average vertical position of our leaves. Add a plot "control" and fill it out to be like Fig. 5.2. We will alter are code to store the current average y coordinate value in the variable called ***leaf-average-y***. After clicking OK to close the dialog box, right-click the plot control and choose select, then resize the plot to an appropriate size. Again right-click and unselect the control.

To calculate the average y position, we first need to add the leaf-average-y variable to our model by using the declaration ***globals [leaf-average-y]*** at the beginning of our code. We then need to calculate the value of the variable during each *go* iteration. The average is just the sum of the leaf y locations divided by the count of the leaves. We can therefore add the following at the end of the *go* procedure (just before the *tick* command).

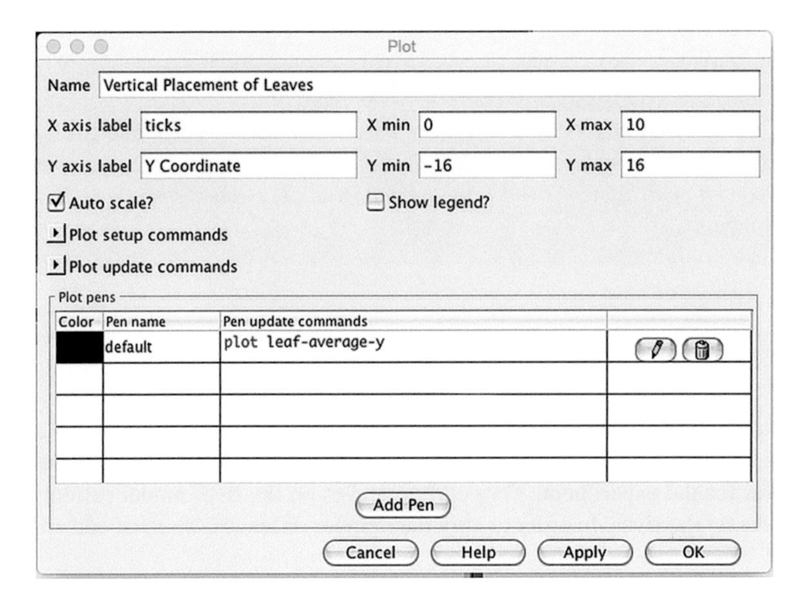

Fig. 5.2 Plot control dialog box

```
set leaf-average-y 0
ask leaves [
 set leaf-average-y leaf-average-y + ycor
]
set leaf-average-y leaf-average-y / count leaves
```

Answer It!

Q05.16: Run the model for at least 200 ticks and screen capture your plot.

Discuss It!

There are many possible improvements to the leaves on a river model. How might you implement an ability to change the number of leaves in the river during the run? With user intervention? Randomly? How might you expand the possible colors of the leaves? How might you make the "water" blue?

5.11 Related Modules

- Module 1: Introduction to Computational Science.
- Module 2: Types of Visualization and Modeling.
- Module 9: Procedures: Performance and Complexity.

Acknowledgement The original version of this module was developed by Dr. Larry Vail.

References

Böhm C, Jacopini G (2011) Structured program theorem. http://en.wikipedia.org/wiki/Structured_program_theorem. Retrieved July 2011

Wilensky U (1999) NetLogo. http://ccl.northwestern.edu/netlogo/. Center for Connected Learning and Computer-Based Modeling, Northwestern University, Evanston

3.11 Related Modules

- Module 1—Introduction to Computational Science
- Module 2—Types of Visualization and Modeling
- Module 7—Simulations, Pseudocode, and Complexity

Solving Equations

6

6.1 Objectives

After completing this module, a student should be able to:

- Describe an analytical solution to an equation.
- Describe a numerical solution to an equation.
- Give examples of when numerical solutions are needed or preferred.
- Describe limitations of iterative numerical solutions.
- Appreciate the need to validate numerical solutions.
- List sources of error in numerical solutions.

6.2 List of Terms

- Analytical solution to an equation
- Convergence
- Error term
- Iterative solution to an equation
- Numerical solution to an equation
- Validation

6.3 Motivation

Mathematics is said to be the language of science. This means it can be used to describe what we observe in science, and then that description can be used to discover unknowns. For example, say we make observation of an object in flight at a number of different times and record its position (time represented by t, and location y). From that data we can use "curve fitting" (see Module 12) to determine the best mathematical model to represent the entire flight path, $f(t) = y$. We can then

K. Brewer, C. Bareiss, *Concise Guide to Computing Foundations*,
DOI 10.1007/978-3-319-29954-9_6

use that model, to determine, for example, where the object would be at the time mid-point, $f(t_{1/2})$. Mathematically we say we need the solution of the equation (function f(t)) when t is equal to $t_{1/2}$. This module discussed the ways equations can be solved using computing technology.

6.4 Discussion

A mathematical equation or set of equations (systems of equations) can be solved if there is the at least the same number of equations as there are unknown variables. Possibly the simplest of solvable equations could be:

$$2 + 2 = x \tag{6.1}$$

which can be solved as we have one equation and one unknown (x). Equation (6.1) also could be rearranged into the format of a linear equation as:

$$x - 2 = 2 \tag{6.2}$$

If we rearrange the equation further,

$$x - 4 = 0 \tag{6.3}$$

we would think of solving for the "root", which is also the solution.

If we have two unknowns, we would need at least two equations, like:

$$2x + y = 1; \quad -3x + 2y = 0 \tag{6.4}$$

Not all sets of equations have unique solutions. Thinking geometrically for a two unknown system, two linear lines can cross at one point (a unique solution), never cross (i.e., the lines are parallel so no solution), or there can be an infinite number of solutions (i.e., the lines overlap).

Discuss It!

Does the equation set in Eq. (6.4) have a unique solution? Can you come up with a pair of equations (with unknowns x, and y) that have no solution? That have an infinite number of solutions?

Mathematical equations (and sets of equations) can be solved in three primary ways: **inspection**, **analytically**, and **numerically** (i.e., by a computer). Inspection is where one can determine the solution by just looking at the equation (or set of equations). Equation (6.1) above is an example of an equation that can be solved by Inspection. Typically, equations used in science and engineering are too complex to solve by inspection.

Try to solve each of the following equations by **inspection**.

$5x + 3 = -2$ (linear equation)
$x^2 + 4x + 2 = 0$ (quadratic equation)
$x^3 - 5x + 1 = 0$ (cubic equation)
$x^4 + 2x^3 + 3x + 4 = 0$ (quartic equation)
$x^5 - x + 1 = 0$ (quintic equation)
$\sin(x) = 0.435$ (trigonometric equation)
$e^x = 25$ (power equation)

You should have experienced the increasing difficulty (quickly reaching "impossible") of trying to correctly solve each of the equations by **inspection**. Fortunately we have more powerful methods.

Solving equations **analytically** means you will use mathematical rules to work the equation(s) into forms that can be easily calculated. Equation (6.4) above is an example, as one would use some sort of mathematical rules (process) to isolate x (to calculate it), and then substitute that value in one of the original equations to then solve for y. There are a number of relatively complex equations in science and engineering have been solved analytically, often using very advanced techniques.

The Symbolab website (https://www.symbolab.com/solver/system-of-equations-calculator) is an easy-to-use service that can solve a variety of equations. Be sure to "show steps" of the solution to see the details of the solution.

Answer It!
Use the website discussed above to find and report the roots (solutions) of the following equations.

Q06.01: $5x + 3 = -2$
Q06.02: $x^2 + 4x + 2 = 0$
Q06.03: $x^3 - 5x + 1 = 0$
Q06.04: $x^4 + 2x^3 + 3x + 4 = 0$
Q06.05: $x^5 - x + 1 = 0$
Q06.06: $\sin(x) = 0.435$
Q06.07: $e^x = 25$

Solving equations **numerically** means you will use the computer to iteratively reach a solution (actually, an acceptably close approximation of the solution). Numerical solutions to a system of equations are covered in more depth in Module 8. There are multiple techniques/methods to solve equations numerically (a thorough coverage of numerical methods can be found at http://nm.mathforcollege.com/) with common methods called Newton's, Lagrange, Spline, Secant, Bisection, Direct, among others.

One free to use numerical solver website (http://www.hvks.com/Numerical/websolver.php) uses P(x) = 0" notation, so any unknown equation will need to be reformulated to equal zero. Also note that to indicate a power, like x^2, you need to enter it as x^2. (It is suggested to turn on "Display Iteration Trail" to see the details of the solution computations.)

The Wolfram Alpha website (http://www.wolframalpha.com) also can be used to solve equations numerically. Wolfram Alpha can be used to solve equations with various methods (by entering "using secant method" or "using newton's method" or "using bisection method" after inputting the equation).

Answer It!

Use the Wolfram Alpha site with the three solution methods (one at a time) to solve each equation. Report on your solution results.

Q06.08: $5x + 3 = -2$

Q06.09: $x^2 + 4x + 2 = 0$

Q06.10: $x^3 - 5x + 1 = 0$

Q06.11: $x^4 + 2x^3 + 3x + 4 = 0$

Q06.12: $x^5 - x + 1 = 0$

Q06.13: $\sin(x) = 0.435$

Q06.14: $e^x = 25$

With the increasingly easy-to-use web services to solve equations numerically, it might be tempting for an engineer or scientist to always rely on that solution method (over inspection or analytically). But recall that numeric solutions are *approximations of the solutions*. Often they are very good, but sometimes not.

Answer It!

Q06.15: Compare your results from the analytic and numeric methods (using just one of the numeric methods used) (i.e. how close is the numeric solution to the anlaytic solution?).

Q06.16: Considering your previous answer, explain why you might and might not want to use numerical solutions for an equation.

6.5 Computing Questions

- For the different integers in your module, what data size(s) do you think were used to represent them? Why?
- For the different floating point numbers in your module, what data size(s) do you think were used to represent them? Why?
- What are the different sources of errors that might be encountered in the computational work done in your module?

6.6 Related Modules

- Module 5: Procedures: Algorithms, Abstraction, And Performance. Introduces concepts of algorithms and their performance when considering numerical methods.
- Module 7: Design and Analysis with Engineering Spreadsheets. Will use iterative solutions to equations.
- Module 8: Flow Analysis. Will extend ideas for solutions of equations to simultaneous systems of equations.
- Module 13: Curve Fitting. Will apply analytic solutions to systems of equations to arrive at best-fit equations for data.

Acknowledgement The original version of this module was developed by Dr. Larry Vail.

Web Resources

http://www.shodor.org/interactivate /activities/EquationSolver/
http://www.wolframalpha.com/input/?i=solving+equations
http://www.hvks.com/Numerical/websolver.php
https://www.symbolab.com/solver/system-of-equations-calculator
http://nm.mathforcollege.com/

Iterative Solutions 7

7.1 Objectives

Students should be able to:

- Create formulas to model data.
- Organize and format data and calculations
- Use formatting tools to improve data readability and visualization.
- Install and use the Excel Goal Seek to identify a solution for a complex model or an optimal parameter for a design or analysis.
- Install and use the Excel Solver to identify a solution for a complex model or an optimal parameter for a design or analysis.

7.2 List of Terms

- Design
- Constraints
- Trial-and-Error
- Iteration
- Parameter

7.3 Motivation

Design "is the process of devising a system, component, or process to meet desired needs." [Haik & Shahin 2010] Even more broadly, "design is to imagine and specify things that don't exist, usually with the aim of bringing them into the world." [Dym & Little 2008] For engineers and scientists, the design process usually involves an analysis of numerical and non-numerical data, in order to determine the appropriateness and safety of the chosen components and systems, and to assess if and how

K. Brewer, C. Bareiss, *Concise Guide to Computing Foundations*,
DOI 10.1007/978-3-319-29954-9_7

well the needs will be met. Sometimes these calculations are simple and straightforward. More often, however, there are multiple interrelated choices that must be made. Computational tools become useful to manage these interdependencies and to facilitate the re-calculations necessary as various design decisions are explored in order to achieve a superior design to meet the given needs.

7.4 An Example: Have a Hang-Up

Consider this basic need: to hang a 45 kg sign. You might suspend it from above by a cable in tension, like this (Fig. 7.1).

To determine the appropriate size for the cable, you have to determine if the stress on the material is within acceptable limits for the cable size. So you could first determine the tensile stress, S_{max}, in the cable based on the force (weight) of the size, relative to the size of the cable.

$$S_{max} = P/A \tag{7.1}$$

where P is the weight of the suspended load and A is the cross-sectional area of the cable.

Recall the weight of a mass is equal to the mass's acceleration due to gravity, or:

$$F = ma => P = mg \tag{7.2}$$

Thus for a load of $m = 45$ kg, $P = m^*g = (45 \text{ kg})(9.8 \text{ m/s}^2) = 440$ N.

The area of a round solid cable (this could also be called a rod) with a circular cross-section of diameter, D, in meters:

$$A = \pi r^2 = \pi(D/2)^2 = \pi D^2/4 \tag{7.3}$$

After calculating the stress for the cable, you would then determine if its allowable stress is greater than the tensile stress (S_{max}) you calculated due to your sign load. If the calculated stress is greater than the allowable stress, you would have to start over with a larger diameter cable.

For a steel cable, the yield stress (also known as the yield strength), S_y, the condition when the material will deform so much that it won't go back to its original

Fig. 7.1 Conceptual diagram of the first example

shape, would be the appropriate limiting stress. The yield stresses (yield strengths) can be found in a material specification table.

But in fact, a designer would probably want to add a **factor of safety, F_s**, – a multiplier to ensure that the design will be well-within the safe limits. Then the limit stress is the yield stress divided by the factor of safety:

$$S_{\text{limit}} = S_{\text{y}}/F_{\text{s}} \tag{7.4}$$

So, the design condition would be that the tensile stress is not greater than the limiting stress:

$$S_{max} \le S_{\text{limit}} \tag{7.5}$$

Answer It!

Q07.01: Look up the yield stress for structural steel using a site such as that found at the Engineering Toolbox website. Report the strength in N/m^2.

Q07.02: If you choose a typical factor of safety of $F_s = 2.0$, what would be the limit stress be? (in N/m^2)

Q07.03: What would be the tensile stress for a 45 kg sign with a cable of diameter 1 inch?

A key aspect of any design process is to find the optimal (or best) design – a design that is most efficient, least expensive, fastest to implement, etc. So for our problem, we could consider just the diameter of the cable to be the one design parameter that we want to be optimal. Going back to our basic equations, we could rearrange and substitute to create a "design condition" equation:

$$\frac{mg}{\pi/_4 D^2} \le \frac{S_y}{F_s} \tag{7.6}$$

and then solve for the ***one*** design parameter, the diameter:

$$D \ge \sqrt{\frac{4mgF_s}{\pi S_y}} \tag{7.7}$$

Answer It!

Q07.04: If you choose a typical factor of safety, $F_s = 2.0$, what would be an adequate size for the structural steel cable to safely hold up a 45 kg sign?

Discuss It!

If you chose another material for the cable, how would that affect our design calculations? Would the cable need to be bigger or smaller? What might be some other design or safety considerations we have not considered?

The rest of this module will use the **Cantilever.xlsx** file found at the textbook website. Open the spreadsheet to the ***HaveAHangUp*** worksheet.

This Excel spreadsheet contains the formulas just described for computing the maximum stress in the cable used to support the sign. The design parameters: $\boldsymbol{m}$, $\boldsymbol{F_s}$, and $\boldsymbol{D}$ can be entered and modified in the yellow-highlighted cells, and the material properties are entered in the blue-highlighted cells. Check out the cell formulas for the computed values such as $\boldsymbol{P}$, $\boldsymbol{A}$, $\boldsymbol{S_{limit}}$, and $\boldsymbol{S_{max}}$. Ensure that the cell formulas reflect the equations above.

Now try entering a design candidate (that is, a trial diameter of the cable) using structural steel for the 45 kg sign. Pick any diameter you desire. Is $\boldsymbol{S_{max}}$ too big or too small when compared with $\boldsymbol{S_{limit}}$?

Answer It!

Q07.05: Now try entering a design candidate (that is, a trial diameter of the cable) using properties consistent with structural steel for the cable, and a 45 kg sign. Pick any diameter you desire. Is $\boldsymbol{S_{max}}$ greater than or less than $\boldsymbol{S_{limit}}$?

Q07.06: What is the ratio of S_{max} to S_{limit}? If it is below 1, your cable diameter shouldn't fail, but could be even smaller (which might make the cable is less expensive or easier to install), if the ratio is above 1, then your cable might not withstand the stress and fail (break).

Discuss It!

A design is considered safe as long as the calculated stress is less than the limit. Generally, we don't want to over design our systems as that costs more. An optimal design is one which the expected maximum stress is closest to our limit. If we calculate the ratio $\boldsymbol{S_{max}/S_{limit}}$, *what would be its optimal value?*

Q07.07: Using a trial-and-error approach, determine the optimal cable diameter.

Q07.08: Calculate the optimal cable diameter for the following similar problem: m = 75 kg, Fs = 2, and by using a structural steel cable.

Discuss It!

How easy was it to find your optimal design for the first (original) problem? Did you find it easier to arrive at the optimum for the second problem?

The trial-and-error approach is slow and repetitive – even when using Excel. But Excel has a built-in data analysis tool (**Goal Seek**) to automate the trial-and-error process. Solver uses a Generalized Reduced Gradient Algorithm that causes it to perform "extensive analyses of the observed outputs and their rates of change as the inputs are varied, to guide the selection of new trial values." (Microsoft 2015)

In the same worksheet, and using all the same inputs as before (except change the mass of the sign to 70 kg), click on the Smax/Slimit cell (J9). From the Data tab, select the What if Analysis... and then select **Goal Seek**. You should see a dialog box like this (Fig. 7.2).

The "Set cell" entry is the cell you had selected (but you can change it if you desire) and that cell should be the one that you are trying to "seek" a particular value for. In our example, we desire to have that cell as close to 1, but not over. Therefore in the "To value" cell, enter a value of 0.99. In the "By changing cell" field, we need to enter the cell that Excel will automatically change until our "Set cell" is at the "To value" value. Set the "By changing cell" to the D input cell (C4). Your dialog box should look like this (Fig. 7.3).

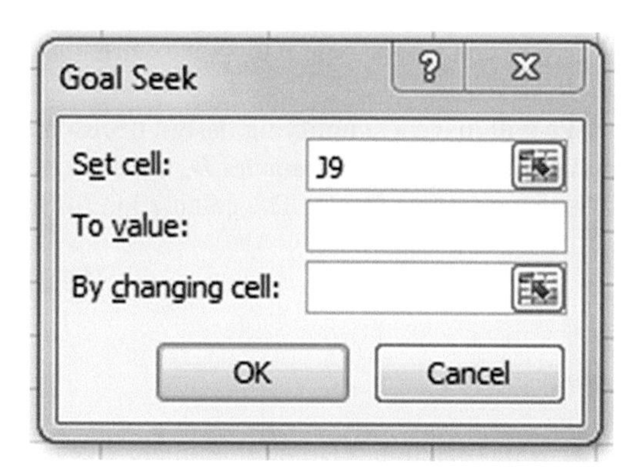

Fig. 7.2 Goal seek dialog box in Microsoft Excel®

Goal Seek
Set cell: J9
To value: 0.99
By changing cell: $C$4
OK
Cancel

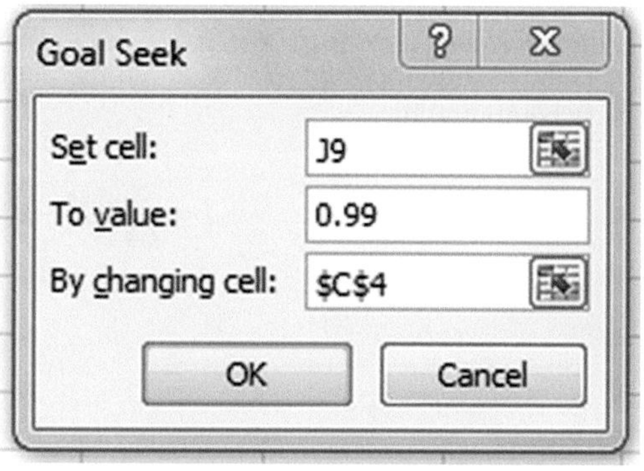

Fig. 7.3 Completed goal seek dialog box in Microsoft Excel®

Now press OK. Excel will keep changing the diameter until the ratio equals 0.99.

Answer It!

Q07.09: What is the diameter of the optimum cable?

Q07.10: Change the cable material to High Strength Steel. What is the optimum cable diameter for this system?

Discuss It!

The Solver tool is a way to automate most trial-and-error calculations. Although commonly used with finding a single optimal design parameter, it can also be helpful when finding solutions to any set of equations where there is only one unknown and the solution is not easily obtained. Think about when the Solver tool could be helpful in your science or engineering discipline.

7.5 Another Design: Out on a Limb...

Another common design choice would be to suspend the sign from a beam mounted on the wall (Fig. 7.4).

For this design the beam would be resisting bending, or in *bending stress*. As a designer, you would want to make sure the bending stress doesn't become a *breaking stress*!

We will make a simplifying design decision and assume that the beam will be a circular tube with outer diameter $\boldsymbol{D_o}$ and wall thickness, $\boldsymbol{b}$. The inner wall diameter $\boldsymbol{D_i}$ is therefore just $\boldsymbol{D_o}$-2$\boldsymbol{b}$. The calculation for the maximum bending stress is:

$$S_{max} = \frac{M y_{max}}{I} \tag{7.8}$$

where

- $\boldsymbol{M}$ is the maximum bending moment and is equal to the weight of the sign, $\boldsymbol{P}$, (as calculated before) times the length of the beam, $\boldsymbol{L}$.
- $\boldsymbol{y_{max}}$ is the maximum distance from the neutral axis (i.e., the center of the tube) to the edge of the beam, which is just $\boldsymbol{D_o}/2$, and
- $\boldsymbol{I}$ is the second moment of area about the neutral axis. For a "thick-walled tube", $\boldsymbol{I}$ is computed by

$$\boldsymbol{I} = \frac{\pi}{64}\left(D_o^4 - D_i^4\right) \tag{7.9}$$

Fig. 7.4 Conceptual diagram of the second example

Using all of the above, we can rewrite our maximum bending stress equation (Eq. 7.8) as:

$$S_{max} = \frac{PL(D_o/2)}{\frac{\pi}{64}\left(D_o^4 - D_i^4\right)} \tag{7.10}$$

which we will now use as our "design condition" equation.

Notice that we now have ***three*** *design parameters*: the tube's (outside) diameter, $\boldsymbol{D_o}$, wall thickness, $\boldsymbol{b}$, and its length, $\boldsymbol{L}$ (i.e., the distance the sign will be hung from the wall.) With just our **one** design condition equation, you cannot solve for the optimal design values for all **three** parameters directly as you did with the previous cable design.

One solution method (actually a pretty good one. . .) is to *guess* a design (a set of values for the three design parameters) that seems like it might work, and check the predicted stress value. You could then make design changes until you find one that you like. If the predicted stress is higher than the allowable limit stress, you could make the tube bigger, shorter, or thicker. If the stress is quite a bit lower, you could try a thinner tube, or allow the sign to be hung further out (if that is desirable.)

Answer It!

Q07.11: If you choose a typical factor of safety, $F_s = 2.0$, what would be the length, outer diameter and wall thickness of a structural steel tube that would safely hold up a 45 kg sign? (Note, we are not asking for an optimum design, just *any* design that would work!)

If your work to answer the prior question seemed like a lot of calculating and re-calculating, you'd be correct. Humans tend to find this tedious, but this is just what computing tools were made for! Furthermore, discovering an error in calculation can send us "back to the drawing board" to start over. Employing a computational tool would let us evaluate a number of possible designs, and hopefully refine and eliminate calculation errors.

Spreadsheet programs, such as Excel, Quattro Pro, and Google Spreadsheets can facilitate a large number of sequential calculations, and are flexible enough to allow a designer to continue to make changes and corrections as the understanding of the problem takes shape.

Open the ***Cantilever.xlsx*** file to the ***OutOnALimb*** worksheet.

This Excel spreadsheet contains the formulas just described for computing the maximum bending stress in the cantilevered tube used to support the sign. The design parameters: $\boldsymbol{m}$, $\boldsymbol{F_s}$, $\boldsymbol{L}$, $\boldsymbol{D_o}$, and $\boldsymbol{b}$, can be entered and modified in the yellow-highlighted cells, and the material properties are entered in the blue-highlighted cells.

Check out the formulas for the computed values such as $\boldsymbol{P}$, $\boldsymbol{D_i}$, $\boldsymbol{I}$, and $\boldsymbol{S_{max}}$. Ensure that the cell formulas reflect the equations above.

Answer It!

Q07.12: Now try entering a design candidate using structural steel for the beam material. Use a beam length of 2.0 m, and choose an outer diameter and wall thickness. Is $\boldsymbol{S_{max}}$ too big or too small when compared with $\boldsymbol{S_{limit}}$?

Q07.13: Now use a beam length of 2.0 m, a F_s of 2, structural steel properties, and a diameter of 0.050 m (that's 5.0 cm, or about 2 inches). How thick must the walls be for $\boldsymbol{S_{max}}$ to be just below $\boldsymbol{S_{limit}}$? (Hint: you can use the Goal Seek tool since there is now only one design parameter that can change.)

Q07.14: For an outside diameter of 0.075 m. How thick must the walls be?

Designers find it useful to compare successive design candidates. To make this easy in Excel, we have reformatted the calculations of the ***OutOnALimb*** worksheet to a table-like format in the ***OutOnALimb2*** worksheet. Enter the same values on Row 6 in this ***OutOnALimb2*** worksheet as you entered on the ***OutOnALimb*** worksheet to ensure the calculations are exactly the same.

On the ***OutOnALimb2*** worksheet, select the cells of your entries and results (that is, cells B6 through P6), then click and hold the small black square in the lower right corner of your selection to "drag and fill" down to make about 12 copied rows. Notice what happened to your input data, and to the formulas for $\boldsymbol{D_i}$ and $\boldsymbol{S_{max}}$.

Answer It!

Q07.15: Enter a different outer diameter in each row (suggestion: start at a diameter of 0.050 m and increase by 0.005 up to 0.100 m). What happened to S_{max} as you increased the diameter of the pipe? Is larger diameter pipe (of the same wall thickness) stronger or weaker?

Again, select the last row of cells and "drag and fill" down to make about 12 more copied rows.

Answer It!

Q07.16: In those new rows, use a constant $\boldsymbol{D_i}$ of 0.100 m and a different wall thickness (***b***) in each new row (suggestion: start at a thickness of 0.005 m and increase by 0.005 up to 0.045 m). What happened to S_{max} as you increased the wall thickness of the pipe? Is a thicker pipe (of the same diameter) stronger or weaker?

Discuss It!

Considering the number of trials (rows) you have created by answering the above, and your results so far, how many rows of different parameter values (for your three design parameters – the tube's (outside) diameter, the wall thickness, and the beam length) would be required to arrive at an optimal design? How might you go about that design process?

This trial-and-error approach using Excel and repeating calculation rows is faster than we could do this by hand, but still doesn't take advantage of *all* the analysis power built-into Excel.

Open the ***OutOnALimb3*** worksheet. This is the essentially the same as ***OutOnALimb2***, but with only one row.

Answer It!

Q07.17: Recall how you used Goal Seek with the cable problem. Use Goal Seek to determine the longest length of the pipe when you are trying to suspend a 50 kg sign using a tube of structural steel that has an outside diameter of 5 cm, a wall thickness of 1 cm, and a safety factor of 2. (Hint: Recall what is the optimum value of the ratio $S_{max}/S_{limit.}$)

Discuss It!

A design could be considered optimal based on a number of different factors – cost being a common and important one. Without knowing the current price of steel, we can substitute weight (mass) for cost, as the price of steel is usually related to weight (mass). Why might that be?

The relative mass of our design will be directed related to the volume of the material in the tube. To calculate the tube volume, recall the equation:

$$\boldsymbol{V} = \boldsymbol{L}\frac{\pi}{4}\left(D_o^2 - D_i^2\right) \tag{7.11}$$

This equation has already been incorporated into the ***OutOnALimb3*** worksheet. Ensure that the cell formulas reflect the equations above.

Answer It!

Q07.18: Using a trial-and-error approach, try 10 different configurations for your three design parameters that keep the maximum stress below the limiting stress (assume a sign mass of 50 kg and a structural steel tube). Keep track of the design configurations. Report your 10 configurations and results (volume and stress ratio).

Q07.19: Consider your results of the previous question and provide a reflection on how optimal your best design is. How many tries do you think you would need?

7.6 The Solution Is at Hand... with Solver

You probably found that it was hard to manage all the parameters and requirements. And it is likely in your 10 tries that you didn't get the "perfect" design, and would not have the patience (or time) to to arrive at a "perfect" design.

This type of problem is common in many fields (STEM and non-STEM) and is called an **optimization problem**. There has been a lot of research and development to devises various ways (and tools) to find the "best" solution for these types of problems. Optimization is covered in detail in Module 14, but for now we will just use the built in Excel tool called ***Solver*** to help find solutions to this type of problem (if it is simple enough). ***Solver*** will determine an optimum value (minimum, maximum, or a target value) for an Objective, by changing Variables, subject to Constraints.

To make our problem work with ***Solver***, let us set an Objective to minimize the volume; by changing the variables of Outside Diameter, Wall Thickness, and Length; with each solution subject to the constraints of making the length at least 2 m, the stress acceptable (ratio as close to 1 as possible), and ensure the wall thickness is less than half the Outside Diameter. We will find that Solver won't let us do that last constraint directly, so instead of setting a wall thickness directly (by having a thickness value), let's set the wall thickness as percentage of tube's radius. The worksheet ***SolutionAtHand*** has this modification.

Discuss It!

Before we use ***Solver***, try your values from your previous best solution to get familiar with this new entry configuration. Ensure the cells calculate all values appropriately. Why is it good science and engineering practice to always test and check computer calculations prior to using?

In Excel, with the ***SolutionAtHand*** worksheet open, from the Data ribbon select the **Solver** tool. (If you don't see the Solver tool in the Analysis section of the Data ribbon, you will need to enable it. On the File tab, click Options, then Add-Ins.

Select Solver Add-in and click the Go button. Check Solver Add-in and click OK. The Solver tool should now be visible in the Data ribbon. Note: The enable process may be different with other versions and platforms.)

Click the ***Solver*** tool and a dialog box will open similar to the following (Fig. 7.5).

Set the Objective to the cell with the value you want to try to maximize/minimize. In our example, that would be the volume (cell M6). Then click "Min" to achieve the minimum of the volume. The Variables will be cells B6:D6, so specify those cells in the appropriate entry box. The Constraints can now be entered. Click the Add button and a new Dialog box appears (Fig. 7.6).

Let us first enter the constraint that the stress ratio needs to be less than or equal to 1. Enter cell R6 in the "Cell Reference" box (remember, you can click the cell on

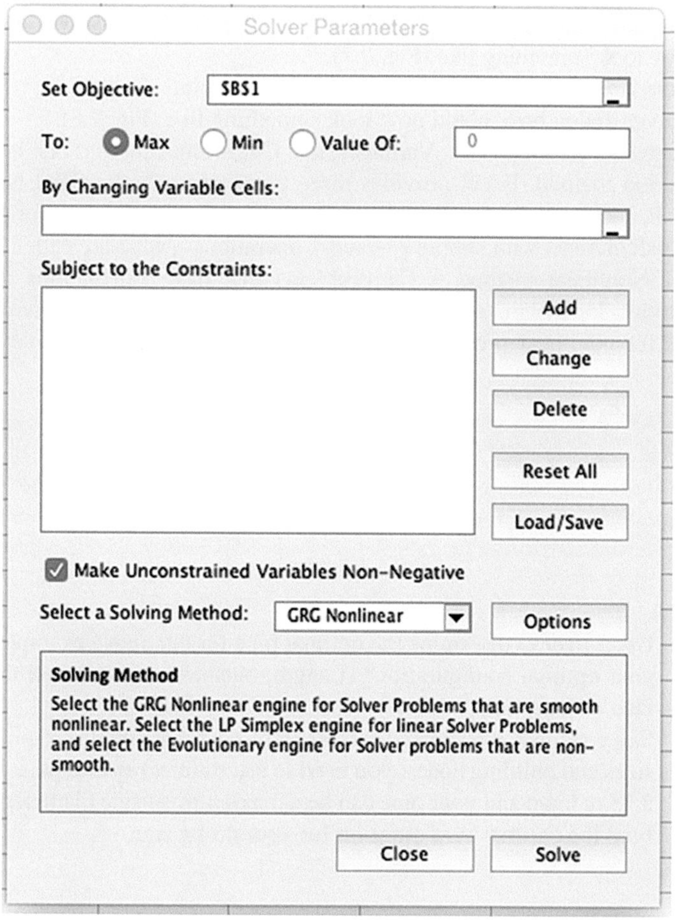

Fig. 7.5 Solver tool dialog box in Microsoft Excel®

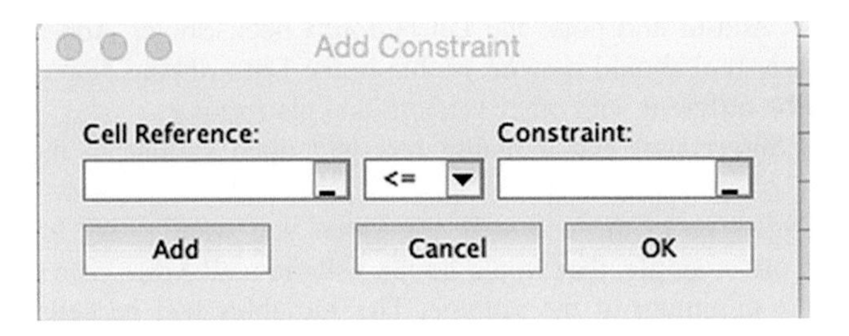

Fig. 7.6 Constraint dialog box in Microsoft Excel®

the worksheet for easy entry). For the "Constraint", enter 1. Then click the Add button. You can now enter more constraints, but lets now go back to the Solver dialog box, so click the Cancel button. (If you accidently click the OK button, you will get an error message, which you can just dismiss.) Your Solver dialog box should now look something like (Fig. 7.7).

Click the Add button to enter the remaining constraints (Table 7.1).

The Solver dialog box should now look something like (Fig. 7.8).

After entering the Objective, Variables, and Constraints, there is one last choice – the solution method. Excel provides three types of methods: GRG Nonlinear, Simplex LP, and Evolutionary. Simply, the Simplex LP method works for problems that have calculations with simple + –/and * operations – what are called "linear". The GRG Nonlinear method is for problems that have calculations involving powers, roots, logarithm, or exponents – that is, nonlinear. The Evolutionary method is for nonlinear problems that can't be solved with the GRG Nonlinear method.

Discuss It!

What solution method would work for our problem? Hint: look at the design equation and supporting equations and determine if they are linear or not.

Answer It!

Q07.20: Use *Solver* to determine the optimal pipe for our above example. Report your optimal configuration? (Length, outside diameter, wall thickness) Don't forget units!

Q07.21: Say you have investigated further and found that you due to pipe supply sizes and building codes, you need to use stainless steel pipe, it has to be 2.75 m long, and your pipe can be a maximum outside diameter of 5 cm. Find the optimal configuration for your 45 kg sign.

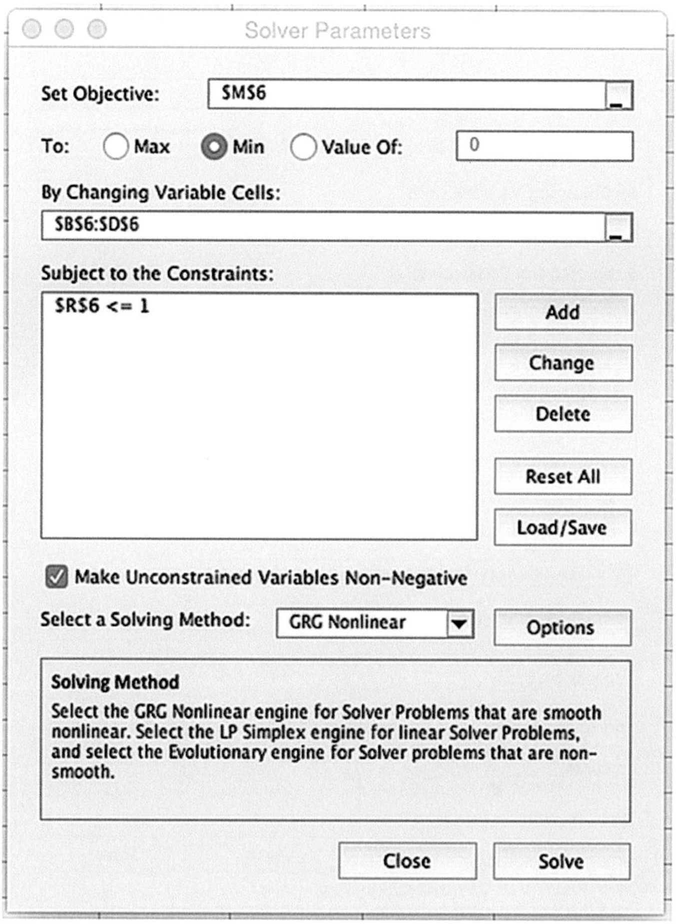

Solver Parameters

Set Objective: $M$6

To: Max Min Value Of: 0

By Changing Variable Cells:

$B$6:$D$6

Subject to the Constraints:

$R$6 <= 1

Add

Change

Delete

Reset All

Load/Save

Make Unconstrained Variables Non-Negative

Select a Solving Method: GRG Nonlinear

Options

Solving Method

Select the GRG Nonlinear engine for Solver Problems that are smooth nonlinear. Select the LP Simplex engine for linear Solver Problems, and select the Evolutionary engine for Solver problems that are non-smooth.

Close

Solve

Fig. 7.7 Completed solver tool dialog box in Microsoft Excel®

Table 7.1 Constraints for example problem

Cell	Logic	Constraint	Comment
L	>=	2	(length of the tube must be at least 2 m)
Do	>=	0.05	(the smallest possible tube is 5 cm in diameter)
Do	<=	0.20	(the largest possible tube is 20 cm in diameter)
b%	>=	0.05	(the thinnest wall is 5 % of the radius)
b%	<=	0.50	(the thickest wall is 50 % of the radius)

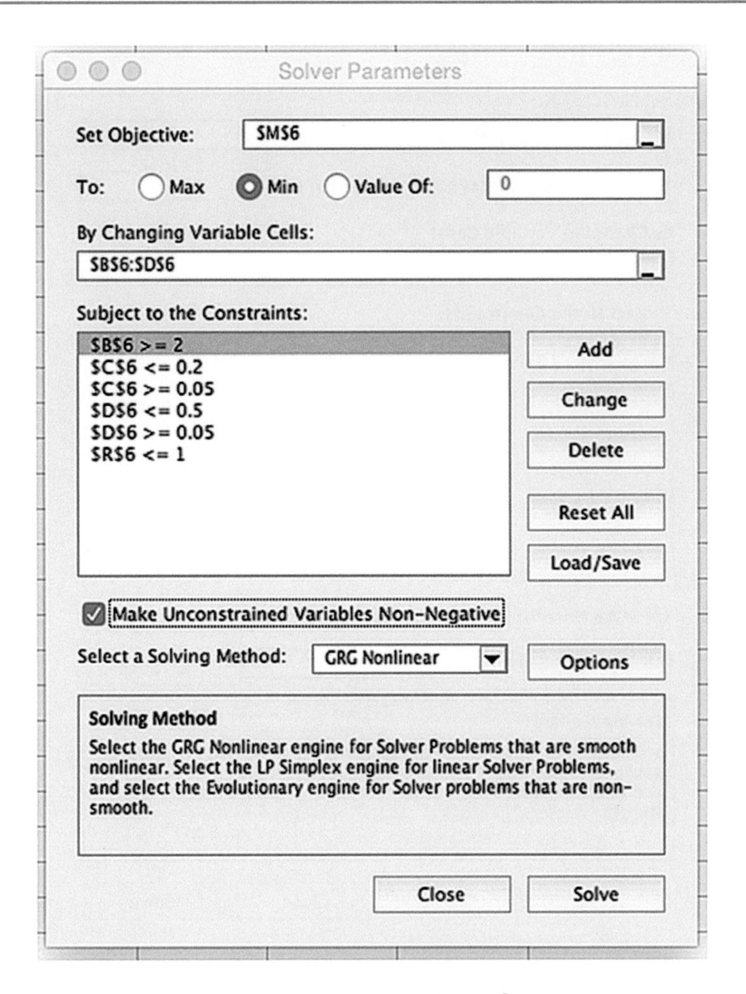

Fig. 7.8 Completed solver tool dialog box in Microsoft Excel®

7.7 But Wait, There's More!

Let's say we decided to hang a 45 kg sign with the stainless steel tube using the optimal design from your answer to the previous question. But much to our horror, when we actually put the sign up, we saw this (Fig. 7.9).

The tube was not horizontal, but bending too much for the comfort of Joe, our client! In fact, Joe has indicated he will only accept a deflection of at most 3 cm from the horizontal.

Fig. 7.9 Conceptual diagram of the third example

What went wrong? We had made sure the tube would not break (fail), but we had neglected to consider the bending – so we need to add that to our calculations. The equation that describes the maximum deflection, ***d***, of a tube is:

$$d = PL^3/3EI \tag{7.12}$$

where ***P***, ***L*** and ***I*** are the same as before, and ***E*** is the Young's Modulus (also known as the Modulus of Elasticity) of the material, which can be found in common online material tables like before.

Open the ***ButWaitTheresMore*** worksheet that has incorporated this equation. And as our practice, ensure that the cell formulas reflect the equations above.

Answer It!

Q07.22: Determine the deflection of the pipe for the above example: 45 kg sign, a tube of Stainless Steel that has an outside diameter of 5 cm, a wall thickness of 3 mm, a safety factor of 2, and a length of 2.75 m. (Hint: These parameters should be close to your optimal solution.

But we know that it does not meet the client's maximum deflection criteria. We need to add the maximum deflection criteria to the ***Solver*** Constraints.

Answer It!

Q07.23: What happens when you use the Solver to determine the optimum (least volume) configuration (length, wall thickness, and outside diameter) for: 45 kg sign, a tube of Stainless Steel that has an outside diameter of between 1 and 6 cm, a wall thickness of between 5 % and 50 % of the radius, a safety factor of 2, a maximum deflection of 3 cm, and a length of 2.75 m.

When trying to answer the above question, the Solver tool should have given you a warning that there was no feasible (possible) solution. This happens when your system has too many or a too restrictive set of constraints. To find a feasible solution, you will need to either loosen or remove constraints, or change some other aspect of your system. Before we change the constraints, let us first determine if a different material will work.

Discuss It!
Before we try different materials in our calculation, look at the given material properties and think about which (if any) materials might be best to minimize the deflection.

Answer It!
Q07.24: Try different materials to see if any of them will allow Solver to find a feasible solution. Report one material that will work.
Q07.25: Now go back to using Stainless Steel and try changing the length constraint to any length at least 2 m long. What is the optimal configuration?
Q07.26: Look at the result from your last calculation. What does the stress ratio tell you about how likely this optimal solution will fail compared to the solution from Q07.22? How has the deflection constraint impacted the safety of your design?

7.8 Integers Are Not Real. . .Numbers

Your optimal design from before likely resulted in some design parameters that were real numbers. Real world materials, however, come in discrete sizes.

Discuss It!
You may have noticed that one of the constraints is "int", which can limit the variation of a cell to only integer values. Think about how you might use that type of constraint with an equation to give you the discrete sizes you need to represent real world materials.

Answer It!
Q07.27: Modify the *IntegersNotReals* worksheet and the constraints in the Solver tool, to limit your size choices to millimeter increments.
Q07.28: For a further challenge, further modify the *IntegersNotReals* worksheet and the constraints in the Solver tool, to limit your length choices to increments of 15 cm.

7.9 Computing Questions

- How reliable were the simulations you used/developed? Why?

- How valid where the simulations you used/developed? Why?
- How did you verify the simulations you used/developed?
- What were the possible sources are errors in your simulation(s)?
- Did your simulation(s) alleviate measure/time/cost and/or ethical issues? Is so, how?
- For the different integers in your module, what data size(s) do you think were used to represent them? Why?
- For the different floating point numbers in your module, what data size(s) do you think were used to represent them? Why?
- What are the different sources of errors that might be encountered in the computational work done in your module?
- For the largest typical size of data used in the area of science studied in your module, how much space is needed? Why?

7.10 Related Modules

- Module 6: Solving Equations. Various techniques for solving equations are introduced.
- Module 8: Solving Sets of Equations. More complex systems of equations with multiple variables are explored and explained.
- Module 13: Curve Fitting. Solving equations to find the best (optimal) equation to match data is explained.
- Module 14: Optimization. More advanced methods to find optimal solutions for complex systems of equations are introduced.

Acknowledgement The original idea for this module was by Dr. Joseph Schroeder.

References

Dym C, Little P (2008) Engineering design: a project based introduction, 3rd edn. Wiley, Hoboken/Chichester
Haik Y, Shahin T (2010) Engineering design process, 2nd edn. CL Engineering, Stamford
Microsoft (2015) https://support.microsoft.com/en-us/kb/82890
http://www.engineeringtoolbox.com/young-modulus-d_417.html

8 Solving Sets of Equations

8.1 Objectives

After completing this module, a student should be able to:

- Describe a boundary value problem.
- Define Finite Difference, boundary condition, domain, iteration, convergence.
- Describe the two most common boundary conditions.
- Appreciate how numerical modeling software works.
- Construct a simple numerical model using Microsoft Excel® to solve a Laplacian system.

8.2 Definitions

- Anisotropic
- Boundary conditions
- Calibration
- Cauchy Boundary Conditions
- Convergence
- Dirichlet Boundary Conditions
- Discritization
- Domain
- Finite difference
- Finite element
- Heterogeneous
- Homogeneous
- Isotropic
- Iteration
- Neumann Boundary Condition
- Sensitivity analysis

K. Brewer, C. Bareiss, *Concise Guide to Computing Foundations*,
DOI 10.1007/978-3-319-29954-9_8

- Steady-state
- Transient
- Uncertainty analysis

8.3 Motivation

As with any scientific endeavor, trying to solve for an unknown variable in a complex (or not-as-complex) system requires a disciplined approach. There are four general steps which are typically taken to solve for the unknown:

1. Examination of physical problem
2. Replacement of physical problem with equivalent mathematical problem
3. Solution of mathematical problem using accepted techniques
4. Interpretation of the mathematical results in terms of the physical problem.

A class of scientific problems centered on determining how something flows through some sort of media will be the focus of this module. (Other, more complex, scientific problems might require a Master's or a Doctorate to fully understand solution techniques!) Flow of water in porous media or flow of heat through some solid media result in similar numerical challenges. This module will specifically use the example of groundwater flow through a non-complex media in this module.

Discuss It!
What kinds of boundary value problems can you think of for your science area?

8.4 Problem Definition

Flow problems can be described by differential equations based on governing (principle) laws of physics. Those differential equations are w.r.t. space and time (x,y,z,t), and form a mathematical model which is known as a boundary-value problem. A full definition of a boundary-value includes:

- size and shape of the flow region (the domain).
- equation of flow within the region (a partial differential equation derived from first principles—principle laws)
- boundary conditions on the domain boundaries.
- initial conditions in the domain (if a transient problem).
- spatial distribution of media property parameters.

- mathematical method of solution: inspection, graphical techniques (flow nets), analytical methods, analog models (electric/resistive, etc.), numerical methods (i.e. modeling—the focus of this module).

Answer It!

Q08.01: Think about a boundary-value problem for flow through a simple (homogeneous, isotropic) earthen dam. (Hint: Think of the problem as a 2D cross-section through the dam with the reservoir on the left and the downstream river on the right.) Would your boundary-value problem be steady-state or transient?

8.5 Boundary Conditions

There are three general types of boundary conditions:

- Type I: Specified Value
  - Known as a Dirichlet boundary condition
  - In hydrogeology, it is known as a Specified Head Boundary.
  - Head can change over time
  - Assumed infinite supply of water
  - Examples: lake; known head value
- Type II: Specified Rate
  - Known as a Neumann boundary condition
  - In hydrogeology, also called a Specified Flux Boundary.
  - Flow rate across the boundary is defined
  - Examples: impermeable boundary; recharge; wells
- Type III: Value Dependent
  - Known as a Cauchy boundary condition
  - In hydrogeology, also called a Head Dependent Boundary.
  - Flux is a function of head difference
  - Because the boundary is dependent on the solution (and the solution is dependent on the boundary), this type of boundary results in a non-linear problem.
  - Examples: lake; streams; rivers

Two other boundary conditions found in hydrogeology are the Free surface boundary (unconfined conditions) and a Seepage face. For this module, only the first two boundary types (Dirichlet and Neumann) will be considered.

In general, it is wise to be careful with over- or under- prescribing boundaries. Too much boundary prescribing may result in an unsolvable system (too constrained). Under-prescribing can result in an unsolvable system (not enough constraint). It is also good practice to not place boundaries near your area of interest.

8.6 Solution Methods

In the various methods of solution listed above, graphical techniques and analog models are considered qualitative—they do not rely on mathematical computations for solutions. Inspection is only applicable for very simple and small problems where no calculations are necessary. The analytical and numerical methods are considered quantitative. Advantages and disadvantages for the common quantitative solution techniques include:

- Analytical/Semi-Analytical
  - Solve simplified versions of the governing equations.
  - "Quick" results since it just involves solving an algebraic equation.
  - Often used to verify results from other solution techniques.
- Finite Difference (FD)
  - Approximates the partial differential equations (pdes) with a set of algebraic equations.
  - Model domain is subdivided into a series of squares (2D) or cubes (3D).
  - Relatively easy to construct and solve, but less robust to handle exceptional geometry domains.
  - Result for each cube is an average value—no assumption of variation within cube.
- Finite Element (FE)
  - Approximates the pde with a series of interpolation functions (basis functions).
  - Grid consists of linked triangles or prisms, which gives flexibility in handling any domain shape.
  - FE are difficult to program, but can be more numerically stable.
  - Result value is defined everywhere in space, but is only calculated at *nodes* (corners of the grid).
  - FD is essentially a special case of FE with a regular grid.

8.7 Numerical Aspects

Consider a simple two-dimensional Laplacian equation which occurs frequently in natural systems (e.g. heat and groundwater flow). In hydrogeology, the equation below represents a two-dimensional steady-state confined aquifer system that has homogeneous and isotropic properties (i.e. the properties are the same in space and direction within the domain).

$$\partial^2 h/\partial x^2 + \partial^2 h/\partial y^2 = 0 \tag{8.1}$$

where $\boldsymbol{x}$ and $\boldsymbol{y}$ are the spatial directions and $\boldsymbol{h}$ is the unknown value you solve for (which for the groundwater flow equation represents head, a measure of energy).

Below are the domain and boundary conditions (Fig. 8.1).

To simplify the system, assume it is square, with no-flow boundaries on all sides except for constant-head boundaries at the top-left corner (with a value of 100) and at the bottom-right corner (with a value of 90). If you choose to discretize the system with a 3×3 regular grid, you would have the following (Fig. 8.2).

This finite difference grid will be used to find the unknown head values at each grid cell, and will assume you are calculating the cell value at the center of each grid. The grid cells are indexed as follows (Fig. 8.3).

Answer It!

Q08.02: What are the indices for the cells that will have no-flow boundaries? What are the indices for cells with constant heads?

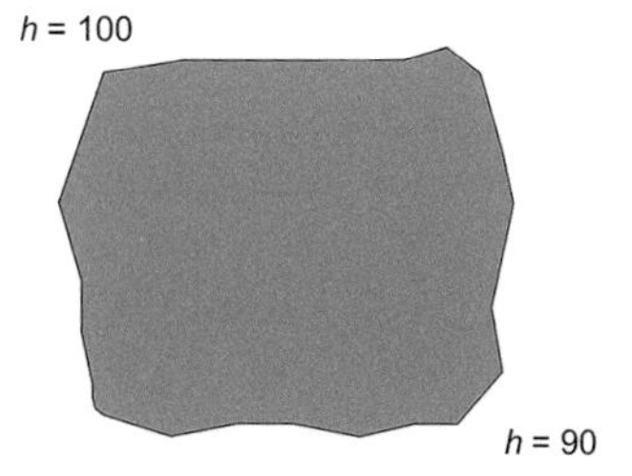

Fig. 8.1 Example domain and boundary conditions

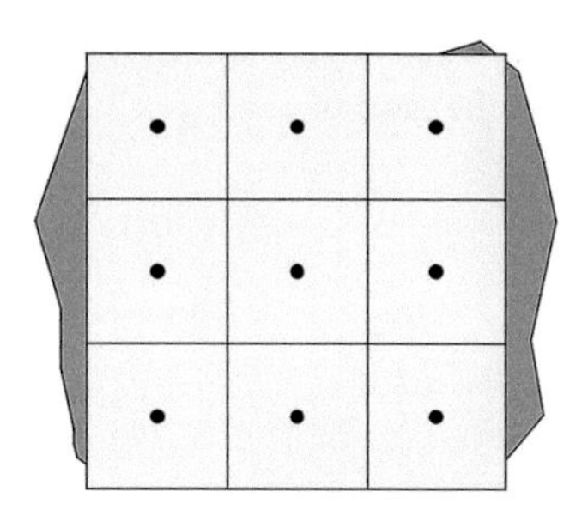

Fig. 8.2 Example domain with 3×3 grid

Fig. 8.3 Example grid indices

$i = 1$

$i = 2$

$i = 3$

$j = 1$ $j = 2$ $j = 3$

Now look back at the flow equation. The finite difference method uses differences to replace the derivatives. The second derivative in the y-direction for a cell at *i,j* can be approximated via a central approximation of two first derivatives:

$$\partial^2 h_{i,j} / \partial y^2 \approx (((h_{i+1,j} - h_{i,j}) / \Delta y) - (h_{i,j} - h_{i\text{-}1,j}) / \Delta y) / \Delta y \qquad (8.2)$$

where $\mathbf{\Delta y}$ is the distance between the cells (cell centers). This equation can be simplified to:

$$\partial^2 h_{i,j} / \partial y^2 \approx (h_{i\text{-}1,j} - 2h_{i,j} + h_{i+1,j}) / (\Delta y)^2 \qquad (8.3)$$

Answer It!

Q08.03: Write the simplified finite difference central approximation in the x-direction.

By substituting in these finite difference approximations for the two second derivatives in the Laplace Equation, you get the following:

$$h_{i-1,j} + h_{i+1,j} + h_{i,j-1} + h_{i,j+1} - 4h_{i,j} = 0 \qquad (8.4)$$

or

$$h\, i,j = (h_{i-1,j} + h_{i+1,j} + h_{i,j-1} + h_{i,j+1}) / 4 \qquad (8.5)$$

which can be just simply said as *the average of my neighbors is my value* (Fig. 8.4). For the case of $i = 2, j = 2$:

$$h_{2,2} = (h_{1,2} + h_{3,2} + h_{2,1} + h_{2,3}) / 4 \qquad (8.6)$$

Discuss It!

What about at non-flow boundary cells, such as i = 2, j = 3? Or i = 2, j = 1?

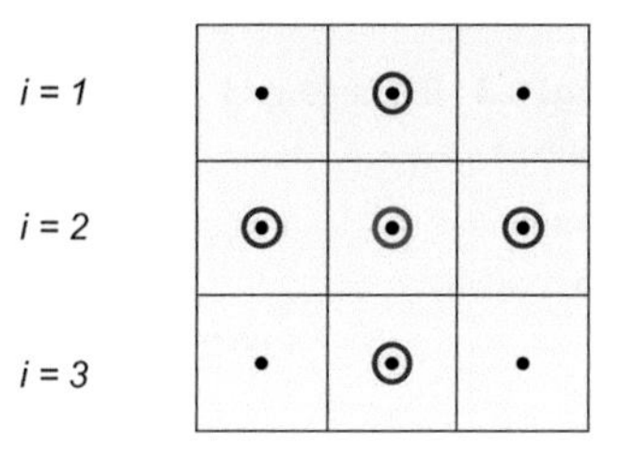

Fig. 8.4 Average of my neighbor cells

Fig. 8.5 Handling boundary cells

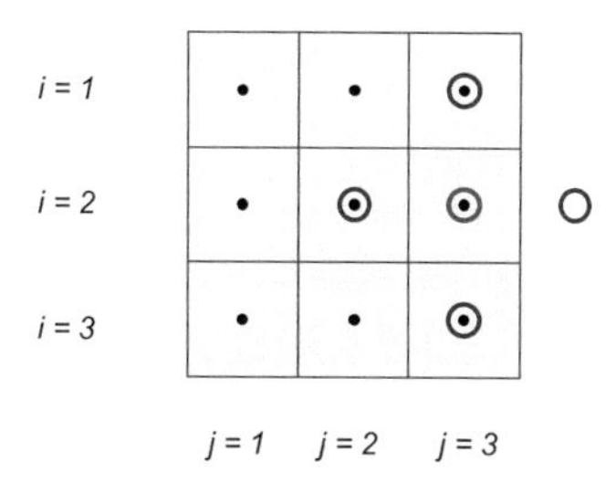

Figure 8.5 shows how the "normal" finite difference approximation would work at $i = 2, j = 1$. Note that one of the values (h at $i = 2, j = 4$) is non-existent. What to do? Since the boundary is no-flow, then it would follow that the energy (h) would be the same (no change in energy, no flow) at the non-existent location as it would be at $i = 2, j = 3$. So the non-existent energy value, $h_{2,4}$ can be thought of as equal to $h_{2,3}$. Substituting this into the typical equation, we get:

$$h_{2,3} = (h_{1,3} + h_{3,3} + h_{2,2} + h_{2,3}) / 4 \tag{8.7}$$

which can be simplified to:

$$h_{2,3} = (h_{1,3} + h_{3,3} + h_{2,2}) / 3 \tag{8.8}$$

Answer It!

Q08.04: Using the same process as that used above, write the simplified finite difference central approximation for the corner cell at $i = 3, j = 1$.

8.8 An Excel Solution

For simple systems with no-flow or constant head boundaries, the above equations can be easily implemented in Microsoft Excel®. But if you were to just enter all the equations in each cell, you would get an Microsoft Excel® warning about circular references— in other words, for example, the answer to cell D4 is dependent on D5, which is dependent on D4. So before you can get Microsoft Excel® to calculate your answer, you need to change some of the calculation settings to allow iteration to a solution.

8.9 Setting Up Excel for Iterative Calculation

Go to the *File* menu and choose *Options*. Go to the *Formulas* tab. Check the *Enable iterative calculation*. Change the maximum number of iterations and maximum change as desired. You may also want to change the workbook calculation to *Manual* and then use the F9 key to start each set of iterations (Fig. 8.6).

On each press of the F9 key, you start another set of iterations. These will stop either when the maximum change in any cell is less than your criteria, or at the maximum number of iterations, whichever is first. You will likely have to press the F9 key multiple times for Excel to iterate to a *converged* solution.

Answer It!

Q08.05: Create an Microsoft Excel® spreadsheet that solves the 3 × 3 grid example problem that was discussed above. Your result should look like Fig. 8.7.

Q08.06: Add color to your solution by using Conditional Formatting / Color Scales on the cells.

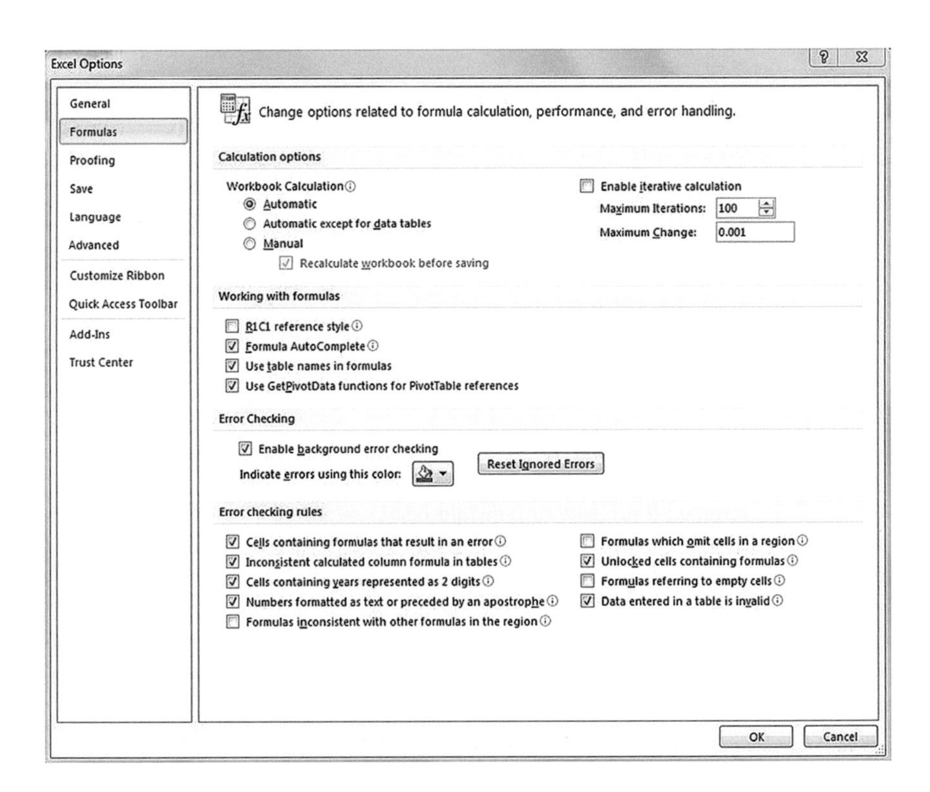

Fig. 8.6 Calculation options in Microsoft Excel®

Fig. 8.7 Solution for Q08.05

100	97	95
97	95	93
95	93	90

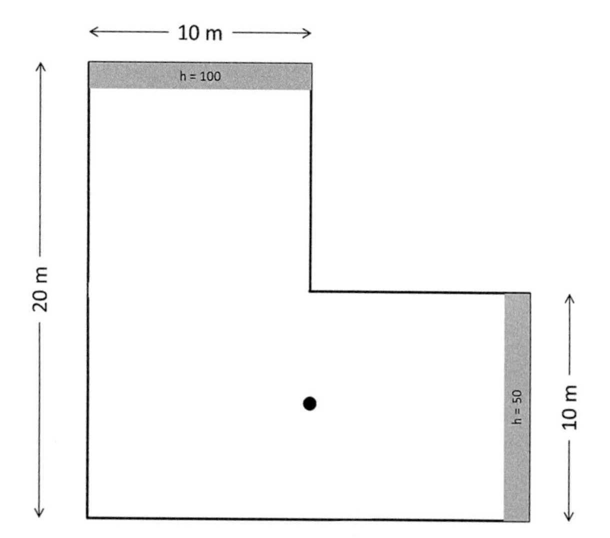

Fig. 8.8 Diagram for Q08.08

Q08.07: Increase the discretization of your problem by solving the same example problem, but with 9×9 cells. Still use only one constant value cell in the upper left and lower right corners of the domain.

Q08.08: Use Excel to solve the boundary value problem shown in Fig. 8.8. Use a grid size ($\Delta x = \Delta y$) of at most 1 m. All head values are in meters. Assume no-flow boundaries for all non-specified head boundaries. For extra credit, have a constant value of 20 m at the dot location.

Discuss It!

So far, you have only considered homogeneous, isotropic systems. If the system was heterogeneous, how could you construct an Excel spreadsheet to account for zones of different flow parameters/properties? The partial differential equation would become:

$$(\partial K / \partial x)(\partial h / \partial x) + (\partial K / \partial y)(\partial h / \partial y) = 0 \qquad (8.9)$$

(with K as the heterogeneous flow property parameter).

8.10 A Matrix Solution

Although the Excel method works well for small problems, it can become cumbersome when used to develop larger models with intricate boundary conditions. Also, modification of the model, once constructed, can sometimes be a bit difficult. Therefore, the common numeric approach is to develop a set of linear equations with the use of various well-developed mathematical (numerical) techniques to arrive at the solution.

Recall that the LaPlace Equation was represented in a finite difference approximation as:

$$h_{i-1,j} + h_{i+1,j} + h_{i,j-1} + h_{i,j+1} - 4h_{i,j} = 0 \tag{8.10}$$

If you were to develop equations for all the cells of our 3 x 3 model, you would have the following set of linear equations (ignoring the constant head boundaries for the moment) (Fig. 8.9).

The equations can be rearranged into a matrix and vector notation format as follows (Fig. 8.10).

$$\begin{aligned}
0 + h_{2,1} + 0 + h_{1,2} - 2h_{1,1} &= 0\\
0 + h_{2,2} + h_{1,1} + h_{1,3} - 3h_{1,2} &= 0\\
0 + h_{2,3} + h_{1,2} + 0 - 2h_{1,3} &= 0\\
h_{1,1} + h_{3,1} + 0 + h_{2,2} - 3h_{2,1} &= 0\\
h_{1,2} + h_{3,2} + h_{2,1} + h_{2,3} - 4h_{2,2} &= 0\\
h_{1,3} + h_{3,3} + h_{2,2} + 0 - 3h_{2,3} &= 0\\
h_{2,1} + 0 + 0 + h_{3,2} - 2h_{3,1} &= 0\\
h_{2,2} + 0 + h_{3,1} + h_{3,3} - 3h_{3,2} &= 0\\
h_{2,3} + 0 + h_{3,2} + 0 - 2h_{3,3} &= 0
\end{aligned}$$

Fig. 8.9 Example system of equations

$$\begin{bmatrix}
-2 & 1 & 0 & 1 & 0 & 0 & 0 & 0 & 0\\
1 & -3 & 1 & 0 & 1 & 0 & 0 & 0 & 0\\
0 & 1 & -2 & 0 & 0 & 1 & 0 & 0 & 0\\
1 & 0 & 0 & -3 & 1 & 0 & 1 & 0 & 0\\
0 & 1 & 0 & 1 & -4 & 1 & 0 & 1 & 0\\
0 & 0 & 1 & 0 & 1 & -3 & 0 & 0 & 1\\
0 & 0 & 0 & 1 & 0 & 0 & -2 & 1 & 0\\
0 & 0 & 0 & 0 & 1 & 0 & 1 & -3 & 1\\
0 & 0 & 0 & 0 & 0 & 1 & 0 & 1 & -2
\end{bmatrix}
\begin{Bmatrix} h_{1,1}\\ h_{1,2}\\ h_{1,3}\\ h_{2,1}\\ h_{2,2}\\ h_{2,3}\\ h_{3,1}\\ h_{3,2}\\ h_{3,3} \end{Bmatrix} = 0$$

Fig. 8.10 Example system of equations in matrix form

$$\begin{bmatrix} 1 & 0 & 0 & 0 & 0 & 0 & 0 & 0 & 0 \\ 1 & -3 & 1 & 0 & 1 & 0 & 0 & 0 & 0 \\ 0 & 1 & -2 & 0 & 0 & 1 & 0 & 0 & 0 \\ 1 & 0 & 0 & -3 & 1 & 0 & 1 & 0 & 0 \\ 0 & 1 & 0 & 1 & -4 & 1 & 0 & 1 & 0 \\ 0 & 0 & 1 & 0 & 1 & -3 & 0 & 0 & 1 \\ 0 & 0 & 0 & 1 & 0 & 0 & -2 & 1 & 0 \\ 0 & 0 & 0 & 0 & 1 & 0 & 1 & -3 & 1 \\ 0 & 0 & 0 & 0 & 0 & 0 & 0 & 0 & 1 \end{bmatrix} \begin{Bmatrix} h_{1,1} \\ h_{1,2} \\ h_{1,3} \\ h_{2,1} \\ h_{2,2} \\ h_{2,3} \\ h_{3,1} \\ h_{3,2} \\ h_{3,3} \end{Bmatrix} = \begin{Bmatrix} 100 \\ 0 \\ 0 \\ 0 \\ 0 \\ 0 \\ 0 \\ 0 \\ 90 \end{Bmatrix}$$

Fig. 8.11 Example of complete system in matrix form

Now add your constant head boundaries. Recall that $h_{1,1} = 100$ and $h_{3,3} = 90$. So the first and last equations in your set should reflect that. Modifying the matrix and right-hand side vector, you get (Fig. 8.11).

Which can be represented as:

$$[A]\{h\} = \{q\} \tag{8.11}$$

where [A] is the matrix of head coefficients, {h} is the unknown head vector, and {q} is the right-hand side terms.

Unfortunately, because the A matrix is often sparse and singular, the inverse of A cannot be determined directly; therefore, iterative methods must be used to solve for the unknown vector of h.

Discuss It!

Using MATLAB, enter in the A matrix and q vector, then determine the unknown h vector by simply inverting A and multiplying by q.

There are many different iterative methods with names like Jacobi, Gauss-Seidel, Strongly Implicit Procedure (SIP), Slice Successive Over Relaxation (SSOR), Preconditioned Conjugate-Gradient (PCG), and even *Multigrid*. Different methods are more efficient for different problems, and scientists who model (who are often just called *modelers*) use knowledge, experience, and intuition to determine the best iterative method to use.

In a simplistic sense, iterating to a solution is like making an initial guess at the value of the unknown h vector and computing the left-hand side and comparing it to the known right-hand side. The difference (i.e. residuals) are then used to make a better guess at the unknown h vector. The left and right-hand sides are again compared and the residuals are calculated. Hopefully the residuals get smaller with each iteration and the system converges to a solution. The iterations end when pre-defined convergence criteria are met or time is up.

8.11 The Modeling Process

A key aspect of modeling is to first understand and clarify the purpose of the modeling effort. Often, it is incorrectly thought of that models are designed to recreate some aspect of reality. No! Models should be constructed to answer specific questions related to complex natural systems.

Discuss It!
Global climate models are being used to help understand the impacts of anthropogenic influences on the atmosphere. What do you think the purpose of the models are, and what are the limits on how the model results should be used?

After the purpose is clearly understood and defined, the modeling can commence. The modeler has many decisions to make from assembling data on properties and boundary conditions, to determining domain discretization and iteration techniques.

Once the model is constructed and converges to a solution, the results need to be checked against known values and parameters modified to allow a match of model results to known values. This is called model *calibration*. When calibrating a model, care is taken to obtain model deviations from known values that are *random* and not biased spatially, temporally, or dependent on the value of the unknown.

Once the model is calibrated, modelers try to find what model results are most sensitive to what parameter changes. This is called *sensitivity analysis*. It is performed on a calibrated model by varying one parameter at a time and evaluating model calibration and other pertinent model results.

After all of the above work is performed, the model can be used to make predictions (possibly incorporating an uncertainty analysis) and documentation of the effort can be created.

Finally, here are a few nuggets of practical wisdom:

- Reality is complex
- A well-defined modeling purpose is necessary
- A model can't model everything
- Model results have inherent uncertainty
- A model should be as simple as possible

Discuss It!
How can you balance the following seemingly conflicting statements?

Reality is complex⇐⇒A model should be as simple as possible

8.12 Computing Questions

- How reliable were the simulations you used/developed? Why?
- How valid where the simulations you used/developed? Why?
- How did you verify the simulations you used/developed?
- What were the possible sources are errors in your simulation(s)?
- Did your simulation(s) alleviate measure/time/cost and/or ethical issues?, If so, how?
- For the different integers in your module, what data size(s) do you think were used to represent them? Why?
- For the different floating point numbers in your module, what data size(s) do you think were used to represent them? Why?
- What are the different sources of errors that might be encountered in the computational work done in your module?
- For the largest typical size of data used in the area of science studied in your module, how much space is needed? Why?

8.13 Related Modules

- Module 3: Data Types: Representation, Abstraction, Limitations. Possible sources of errors.
- Module 6: Solving Equations. Various techniques to solve equations are explored.

Further Study

This module's discussion was focused on the Finite Difference Method. Another approach is to use the more flexible (but mathematically more complex) Finite Element Method. The following resources may be a good starting point to further study.

- Akin, Application and implementation of finite element methods, Academic Press, 1982
- Celia MA, Gray WG (1992) Numerical methods for differential equations. Prentice Hall, New Jersey
- Desai and Abel (1972) Introduction to the finite element method. Van Nostrand Reinhold Company
- Kwon and Bang, The finite element method using MATLAB, CRC Press, 1997
- Wang HF, Anderson MP (1982) Introduction to groundwater modeling: finite difference and finite element methods. Freeman, New York

Procedures: Performance and Complexity 9

9.1 Objectives

After completing this module, a student should be able to:

- List computing environment components affecting performance.
- List and rank the standard Big-Oh computational complexity functions.
- Determine the time complexity function of an algorithm.
- Classify an algorithm by its standard Big-Oh function.

9.2 Definitions

- Big-Oh classes
- Dominant term
- Performance
- Procedure
- Time complexity function

9.3 Motivation

This module addresses the computing concepts of *procedures* and their *performance* in today's computing environments.

You will learn about the execution time of procedures and the classification of their computational complexity by using the NetLogo (Wilensky 1999) system and programming language. The objective is for you to become familiar with the concepts of a program's execution time, performance, and its computational complexity.

K. Brewer, C. Bareiss, *Concise Guide to Computing Foundations*,
DOI 10.1007/978-3-319-29954-9_9

9.4 Simulation Model Performance

When using a computer system to do any number of things, one thing you want is good *performance*. Whether you are playing a network game, catching up with friends on Facebook, or downloading the latest software application (i.e. app), good performance is important to everyone. You just don't want to wait, so it is critical that the computer system be able to complete the requested *work* in as little *time* as possible. Computational science simulation models are another type of computer app.

The time for a system to complete some work is inversely proportional to the performance. That means if the time a system takes to do something increases, you say the performance decreases. Restated, if the performance of a system increases, the time to do something decreases. Systems that take more time have lower performance and ones that take less time have higher performance.

Answer It!

Q09.01. Define *performance* in your own words.

Discuss It!

Consider a common application like Microsoft Excel. Name the different components affecting the performance of the application (i.e. how long it takes for certain commands to complete).

Some of the following components make up the computing environment and contribute to the performance of the Facebook website. A browser (e.g. Internet Explorer, Safari, Chrome, Firefox) is running on a computer you are working on. Different browser *software* may be faster than others. Operating system software (e.g. Windows, Mac OS X, Linux, Android, iOS) and different versions (e.g. Windows XP, Vista, 7) also affect performance. The computer *hardware* you use affects performance. Wouldn't it be great to have a fast new computer processor and graphics display? However, you need to realize that the overall system performance is also affected by the *network communication* speeds and algorithms used. Performance will also depend on the amount of *data* that must be stored to or retrieved from *storage*.

Google has made famous the concept of *Cloud Computing*. When you need to search the Internet, you can ask the computing components Google has available in *The Cloud* to perform the search and return the results quickly. This means you don't have to wait long and your computer performance appears faster with the help of other computers on the Internet in the Cloud. You can also store and retrieve pictures to Cloud storage using Google's Photos Web app.

High performance computational science applications can benefit by getting assistance from supercomputers accessible via the Cloud. However, high

Performance Computing (HPC) and supercomputing applications are beyond the scope of this module.

Answer It!

Q09.02. List at least three computing environment components affecting performance.

So performance of a computer system depends on how much work needs to be done and how much time it takes to do that work. The work of computers is to move data from where it is to where it needs to be as well as processing that data. The time to move data depends on how much there is, the speed of the storage devices, and the speed of the communication networks.

Hardware components are critical to good system performance. However, now you want to focus on the time complexity of some simplified algorithms that create or change data in some way.

9.5 Example of Computational Complexity: Tick Marks

You may be familiar with making tick marks when counting something like votes. You usually make a mark for each vote and group them by fives so it will be easier to total them later by counting the groups of five and multiplying by 5. For this example, you will be the computer and the algorithm is placing ticks on a paper. The ticks represent the work to be done (Fig. 9.1).

Answer It!

Q09.03. Take a blank piece of paper and number the first line 0. Make a note of the time and then make 1 tick mark on line 0 of the paper. On the next line, line 1, make twice the number of ticks (i.e. 2) as the number of marks on the previous line (i.e. line 0 in this case). Continue to number lines up to 9. Each line should have twice the number of ticks as the previous line. Approximately how long did it take you to make the ticks on lines 0–9?

Discuss It!

What if you doubled the number of lines to 20 in this problem? How long would it take you?

Since 20 lines are twice as many as 10 lines, you might think it will take you twice as long, but you would be wrong. It will take about 1000 times longer. You may need more room on your paper, too. Don't try it! You have many more important things to think about and to do.

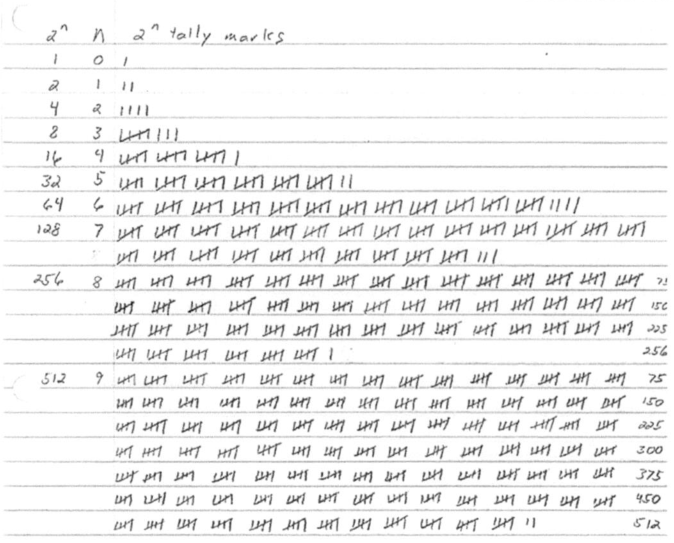

Fig. 9.1 Tick marking example

Let's do some thinking to analyze the amount of work the computer must do using this algorithm. The number of lines is the input to the algorithm. The time is related to the number of ticks you must make on the paper. What you want is a function that when you give it the number of lines to write, it gives you the number of ticks you must make. If you know the number of lines and how long it takes to make a tick mark, you can calculate how long it will take to make all the marks before doing it.

Each subsequent line has twice the number of ticks. You repeatedly multiply the number of ticks by two, and repeated multiplication is represented by exponentiation. Line 0 has one tick. Line 1 has two ticks (i.e. 2^1). Line 2 has four ticks (i.e. 2^2). Line 3 has eight ticks (i.e. 2^3).

Answer It!

Q09.04. How many ticks would be on line 9 for our example?

Q09.05. How many ticks are on line number *n-1*? Line *n*?

If we examine some of the total ticks on particular lines, we might be able to determine how a general function for total ticks could be structured, given the number of lines. Given 1 line, there is a total of one tick. Given 2 lines, there are a total of $1+2=3$ ticks. Note that 3 is equal to 2^2-1. Given 3 lines, there are $3+4=7$ ticks (i.e. 2^3-1). Given 4 lines, there are $7+8=15$ (i.e. 2^4-1). And so on.

Answer It!

Q09.06. What is the total number of ticks for lines 0 through 9 (10 lines)? For 20 lines?

Q09.07. Generalizing your understanding, write an equation that will calculate the number of ticks that must be marked for *n* number of lines.

By now we have a function that will determine the total number of ticks that must be made when given the number of lines. If you know how long it takes to make 1 tick, all you have to do is multiply that time by the number of ticks and you will know how much total time it will take. This is a time function for the algorithm. If you can make on average 1 tick mark per second, using our function we could calculate that it would take 1023 s to complete 10 lines of ticks.

Answer It!

Q09.08. How many minutes is 1023 s? (show your conversion calculation)

You may be thinking, all this has been silly because computer systems are so fast today that it doesn't matter. But let us do some calculations and see. To make the calculations a little easier, assume that a computer writes a tick mark in 10^{-9} s (i.e. 1 billionth of a second or 1 ns).

Answer It!

Q09.09. How many ticks could this hypothetical computer make in 1 s? (Give your answer in scientific notation.)

Using the speed of this hypothetical computer, let us do some time calculations for this algorithm to run given some different numbers of lines. Show your answers to the questions below using scientific notation.

Answer It!

Q09.10. How long will it take our computer to complete: 10 lines? 20 lines? 50 lines? 100 lines? Report your answer for each in seconds and in years.

Your answer to the above question for 10 lines should indicate that it would take about one millionth of a second which is almost instantaneous - imperceptible to humans. However, to complete 100 lines at this rate should complete in approximately 40,000 billion years. Astronomers and geologists estimate the age of the earth to be about 4 billion years, which means the computer would take 10,000 times longer than from earth's creation until now to write the ticks for 100 lines.

If you tried to compute how long many ticks would be required for 333 lines, your scientific calculator will likely exceeded its limit and displayed an error. Exponential functions "grow" fast. (For more on number representation and limits, see Module 3: Data Types: Representation, Abstraction, Limitations).

The exponential nature this tick mark algorithm makes the time to run the algorithm impractical except for very small numbers of lines. Thus, it is said to be computationally complex.

9.6 Another Example of Computational Complexity: Color a Square of Patches in NetLogo

Start NetLogo. On the *Interface* tab click the *Settings* button. Note that the NetLogo world is a rectangular grid of rectangular patches. By default, it is 33 patches wide (i.e. −16 to 16) and 33 patches high. Click the *OK* button to accept the default dimensions and return to the *Interface* tab.

Answer It!

Q09.11. How many total patches are there in the default NetLogo world?

Now, make sure the *view updates* box is checked and the *continuous* option is selected. Then move the *speed* slider to about ¼ scale (*slower*). This may have to be adjusted so you can see individual patches change.

Click in the NetLogo command center and press tab until you see "patches>" which means that you are communicating with the *patches* agents. Enter the command ***set pcolor blue***. This should set all the patches in the world to blue. The patches should be changing one at a time at random until the whole world is blue. If the world appears to change color all at once, adjust the speed slider to slow down the update. Try setting the patch color to another color (e.g. red, green, black) and watch the world change color one patch at a time. If the update goes too fast or too slow, move the speed slider till the update completes in a reasonable amount of time.

The work the computer is doing is different from the previous example, but think about the similarities. The procedure that is running must set the color of each patch in the world. Setting the color of a patch is similar to making a tick mark. The number of patches on a side relates to the amount of work that must be done and so is similar to the number of lines to make. Thus, we can construct an equation that relates the input (number of patches on a side) to the amount of work that must be done, and eventually how long it will take.

Answer It!

Q09.12. Assume it takes 1 s to set the color of one patch. How long will it take to change the color of all patches in a square NetLogo world with 33 patches on a side? In seconds? In minutes?

You should now have an idea of a function that relates the amount of work given the number of patches on the side of the world.

Answer It!

Q09.13. How long will it take to change the color of all patches in a world of size 201? In seconds? In minutes?

Q09.14. How long will it take to change the color of all patches in a world of size 401? In seconds? In minutes?

Discuss It!

In NetLogo, change the Settings to **max-pxcor** *100 and* **max-pycor** *100 and* **Patch size** *2, then click OK. Run your command to* **set pcolor** *to some color for all the patches. This world has 201 patches on each side. (Don't change the speed!) Try it with 401 patches on each side. Compare your calculated time to the actual time.*

Module 14: Data Organization and Analysis will demonstrate that some spatial data is stored as a grid of squares. As you increase the size of the grid, the amount of work and also the run time will increase by the square of the increase. Thus, the size of data "squares" will dramatically affect the amount of work the computer system must do and therefore how much time it takes.

Discuss It!

Think about how the work would increase when representing some sort of spatial data for Illinois at a grid size of 10 × 10 miles vs. 1 × 1 mile (equivalent to the typical spacing of the main rural roads).

Answer It!

Q09.15. Which time function grows faster as the input increases, the exponential 2^n or the square n^2?

By analyzing an algorithm and with practice, you should be able to come up with a function that calculates the time to run, given the input size. It may not calculate the exact time but a value that is proportional to the run time. The rate at which a time function grows as the input size increases affects how computationally complex the algorithm is. Faster growth means more computationally complex. Some algorithms by nature will be impractical to compute in reasonable times for larger sizes of input. It is important to discover these impracticalities before attempting to compute something that will take too long.

9.7 Example of Computational Complexity: Merge Sort

The previous two examples developed *time complexity functions* for two different types of algorithms. Counting with tick marks ended up being proportional to 2^n. Coloring the patches in a square world ended up being proportional to n^2.

Now we will analyze the computational complexity of one more algorithm that will put a list of items in order – *merge sort*. Sorting is a powerful tool for making sense of data. Merge sort is one of several different algorithms that can sort a list. Our goal again is to develop a function that will calculate the amount of work and estimate the time to sort a list given the number of items in the list.

The start of the algorithm puts just one item in each "group" (essentially each item in the list is in its own group). The algorithm then merges each adjacent group into new groups that are twice as long, and simultaneously arranges the items in each group in correct order. This merging and sorting continues until the entire list is sorted (Fig. 9.2).

To merge two groups of length n into one group of length $2n$:

1. Compare the first items of the two groups to each other
2. Put the smallest item of the two shorter groups in a new sorted group that will be twice as long
3. Remove that item from its group
4. Repeat until one of the source groups is empty
5. Place all of the remaining items from the non-empty group on the end of the new sorted group. (Notice that the new sorted group is twice as long as the two sorted groups being merged.)

Repeat this merge for each pair of groups until there is only one group left. This final group is the original set of numbers in sorted order.

Look at the NetLogo MergeSort model (Wilensky 2005). It not only sorts a list of numbers using the MergeSort algorithm, but helps to understand the algorithm by visualizing the work as it progresses. Click the *setup* button and you should see 16 unsorted random positive integers. Each number in row 0 is a different color to indicate that the length of the sorted subgroups is 1.

Click the *step (1 item)* button 16 times watching how the merge part of the algorithm takes adjacent pairs of numbers and merges them into sorted groups of length 2. Row 1 should have 8 sorted groups of length 2.

Keep stepping the model until you finish row 2.

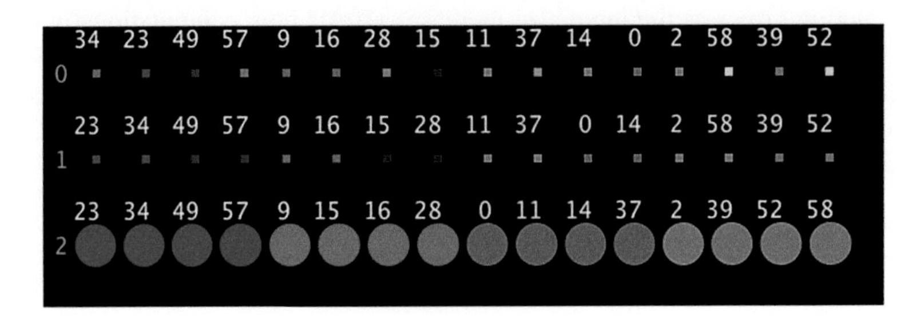

Fig. 9.2 The MergeSort algorithm from NetLogo

Answer It!

Q09.16. How many times do you have to click *step (1 item)* to go through a row?

Q09.17. How many groups are there on row 2?

Keep stepping the model until all 16 numbers are sorted. Run the model several times until you understand how the algorithm works.

Answer It!

Q09.18. How many rows did you have to build before the list was sorted?

Q09.19. How many times do you have to *step (1 item)* before 16 numbers are sorted?

Q09.20. How many times would you have to *step (1 item)* before 32 numbers are sorted?

Q09.21. How many rows would be built for a list of 1024 numbers?

Q09.22. How many times would you have to *step (1 item)* to sort 1024 numbers?

The number of steps required to sort n items using this algorithm is the ceiling (i.e. next larger integer) of logarithm (base 2) of n (i.e. log_2n). Each step requires n comparisons between the numbers. Therefore, the time it takes for the algorithm to run is about $n \log n$. That means that the time complexity of the algorithm grows proportionally to or "on the order of" $n \log n$.

Answer It!

Q09.23. Which time function grows faster as the input increases, $n \log n$ or n^2?

9.8 Standard Big-Oh Function Classifications for Comparing Algorithms

By now, you have had a little practice in analyzing algorithms. You have analyzed three algorithms and developed time complexity functions for each one. The time to complete each of these tasks depends on several different factors in the computing environment. You made some educated assumptions about how long it might take to make a single mark, color a single patch, or move a single number. You were able to predict an estimate for the total time to do the work by multiplying that number by the number of times the algorithm had to repeat that work. Time functions for each of the three examples were proportional to the number of times the algorithm repeated a simple task. Respectively they were proportional to 2^n, n^2, and $n \log_2 n$. To compare the performance of algorithms, you are interested in the rate of growth for their run times compared to each other.

Answer It!

Q09.24. Use a graphing tool to compare the relative run times for the three algorithms (2^n, n^2, and $n\ log_2 n$) for values of n from 0 to 10.

Q09.25. Based on the size of the input *n*, put the three algorithms in order from fastest (i.e. smallest run time) to slowest (i.e. longest run time). Explain why you think so based on your comparison graph.

Time complexity functions developed for algorithms may have different tasks completed one after the other. They will have different terms added together. For example $T(n) = n^2 + 2n + 5$.

Answer It!

Q09.26. Use a graphing tool to compare the size of the three terms in the time complexity function $T(n) = n^2 + 2n + 5$ for n from 0 to 10.

Q09.27. Which of the terms in $T(n) = n^2 + 2n + 5$ grows the fastest as n increases?

The term in a time complexity function that grows the fastest is referred to as the *dominant term*. It is typically much larger than the rest of the terms in the time function. That means that time function behavior can be approximated by the dominant term for comparisons.

Answer It!

Q09.28. Which is the dominant term in $T(n) = n^2 + 2n + 5$?

The list of seven standard complexity functions (n, 1, n^2, log n, 2^n, n!, n log n) can be used to simplify and classify the computational complexity of algorithms. They are called Big-Oh functions and are denoted with the capital letter O. Big-Oh functions ("on the order of") give the approximate (i.e. ball park) complexity of an algorithm.

For example, for an algorithm with time complexity function $T(n) = 20n \log n + 300 + 5n$, the Big-Oh function is O(n) because the dominant term is 5n.

Answer It!

Q09.29. List in increasing order (i.e. from smallest to largest) the seven standard Big-Oh computational time complexity functions (e.g. O(n)).

Q09.30. What is the dominant term of an algorithm with time complexity function $T(n) = 10n^2 + 25 + 2n$? What is the Big-Oh time complexity function for the algorithm?

Discuss It!

Think back over your experience with various procedures and their time complexity and therefore impact on performance. Classify an algorithm by its standard Big-Oh computational complexity function. State/summarize/briefly describe your algorithm. What is the time complexity function for your algorithm? Identify the dominant term for your algorithm. Using the dominant term, simplify the complexity function to the closest standard Big-Oh function that characterizes the run time for the procedure based on the input data value.

9.9 Related Modules

- Module 1: Introduction to Computational Science.
- Module 3: Data Types: Representation, Abstraction, Limitations.
- Module 5: Procedures: Algorithms and Abstraction.
- Module 10: Self-Defining Data: Compression, XML, and Databases.
- Module 13: Optimization.
- Module 14: Data Organization and Analysis.

Acknowledgement The original version of this module was developed by Dr. Larry Vail.

References

Wilensky U (1999) NetLogo. http://ccl.northwestern.edu/netlogo/. Center for Connected Learning and Computer-Based Modeling, Northwestern University, Evanston, IL

Wilensky U (2005) NetLogo merge sort model. http://ccl.northwestern.edu/netlogo/models/MergeSort. Center for Connected Learning and Computer-Based Modeling, Northwestern University, Evanston, IL

http://bigocheatsheet.com/

Self-Defining Data: Compression, XML and Databases

10

10.1 Objectives

After completing this module, a student should be able to:

- Appreciate the power of self-defining data.
- Know when to use different types of compression.
- Read simple XML documents.
- Make simple queries in a relational database.
- Appreciate the demands of efficiency on a DBMS.
- Understand the purpose of data warehousing and mining.

10.2 Definitions

- Compression
- Database
- DBMS
- Lossless compression
- Lossy compression
- Markup language
- Query
- Relation
- Self-defining

10.3 Motivation

There are many areas of study in modern science and engineering that deal with significant amounts of data. For example, recent (at the time of this writing, Summer 2016) star catalogs include around one billion different stars with multiple

K. Brewer, C. Bareiss, *Concise Guide to Computing Foundations*,
DOI 10.1007/978-3-319-29954-9_10

attributes (location, magnitude, etc.) for each. Another example is the recent effort at taxonomic revision of Diplodocidae, resulting in the validation (by their approach) of the genus Brontosaurus (Tschopp et al. 2015). That effort required analysis, classification, and management of 477 morphological characteristics for 81 different "operational taxonomic units".

In this module you will learn how self-defining data in computing can assist scientists and engineers with managing and analyzing large (and small) data. Three different types of self-defining data will be addressed: compressed data, xml documents, and databases. In fact, this very document can be thought of as self-defining because it describes itself. When data is self-describing, the application software does not need to know about the data structure. It can ask the data itself how it is structured and then adapt to it. This allows the structure of the data to change (when needed) without necessarily having to change the software.

10.4 Self-Defining Type 1: Compression

Compression is used to lessen the amount of space needed to store information. Information is compressed while it is being stored and then uncompressed when it is needed. When working with compression, the first thing stored in the file is the information necessary to tell programs how to compress/uncompress the data. Therefore, compression is our first form of self-defining data. There are two major categories of compression: *lossy* and *lossless*. With lossy compression, when the information is uncompressed, some of the original information is lost; however, with a lossless compression, none of the original information will ever be lost. The lossy compression usually will save more space at the cost of information loss.

When doing lossy compression, the goal is to lose information that is not considered important.

Examples of different types of compression of images can be found in many locations on the Internet. Look at each of these and see when you can notice the differences. Also pay attention to the size of the files.

- https://commons.wikimedia.org/wiki/File:Quality_comparison_jpg_vs_saveforweb.jpg
- http://users.wfu.edu/matthews/misc/graphics/formats/formats/
- http://athleticaid.com/yaquinapress/image/

Although images are a common application for compression, other data can be compressed.

Answer It!

Q10.01: What is the range of hearing for a normal human?

Q10.02: What resolution (i.e. how many pixels per inch) can a normal human distinguish?

Q10.03: How many different colors can the human eye distinguish?
Q10.04: Challenge: Answer the same questions for a different animal.
Q10.05: What is a reason for storing information beyond what the human can sense?
Q10.06: What is a reason for not storing information beyond what the human can sense?

The computer (with its sensors) can be designed to identify much more information (types and range) than the human body can. Sometimes this information may not need to be stored. This is where a lossy compression would be used.

However, there are times when information should not be lost (no matter what). For example, a program should never lose even one line of code. If a program loses even one bit of information, it will probably cease to work. Large arrays of simulation data outputs, and raw data from sensor systems should never use lossy compression. Sometimes even things like pictures and audio should not be compressed using lossy algorithms; for example, evidence in criminal proceedings, and gridded geographic information (aerial photographs or LIDAR data).

Sometimes the colors of an image (or visual representation of gridded data) can be adapted to show hidden knowledge. It might be possible (because of shadows, etc.) for the human eye to not be able to read text but for the difference to be stored. In addition, when working in the sciences, there are many types of information used that are outside the range of human perception.

Discuss It!
If a color map is changed, a previously unseen difference/feature in an image may become visible to the human eye. Can you find an example on the web?

Answer It!
Q10.07: Give two different examples in an area of science that uses information that is outside the range of human perception.
Q10.08: What type of compression is used in.mp3 files -> lossy or lossless compression? Why do hard-core, expert musicians often prefer the vinyl format over.mp3?
Q10.09: What type of compression is used when "zipping" a file?
Q10.10: What is some data in an area of science that could be compressed with a lossy compression?
Q10.11: What is some data in an area of science that should only be compressed with a lossless compression?

10.5 Self-Defining Type 2: XML

Sometimes the structure of the information does not fit one of the 10 or so standard structures (including stacks, queues, trees, and graphs). People representing information often want to define their own structure. This is what XML is designed to allow. XML (eXtendable Markup Language) allows people to define their own data structure and is used in many different places (including the recent versions of Microsoft Office). When using XML, it is important that either the program knows and follows the structure defined in the document, or can determine the structure and adapt to it. Just because a program can use a biology XML document does not mean it can use a chemical XML document. Each document type needs to be understood by the program using it. Side note: XML documents can easily be stored in databases (which is covered in the next section).

Answer It!
One example of a use of XML in science is the storing of the periodic table. One location of it can be found at: http://www.satimage.fr/software/samples/allelements.xml. Read the document and find the information associated with your favorite element.

Q10.12: For the periodic table in XML, give two details associated with your favorite element that you did not have previously memorized.
Q10.13: Pretend you have discovered a new element (be creative). Add information about that element.
Q10.14: Find an XML document that is used in your area of science. What does it describe? What are some of the characteristics covered by the document? For one example, list the data associated with those characteristics.
Q10.15: How hard is it to understand an XML document in your area?

10.6 Self-Defining Type 3: Databases

One common way of defining a database is a *self-describing collection of integrated records*. There are a number of different ways to "integrate" the records with the most common being called a relational model. In the relational model, information is stored in relations (tables). The information is integrated by sharing common columns. A small example can be seen below (Tables 10.1, 10.2, 10.3 and 10.4).

How would you find out how much it costs to creating baking soda? This demonstrates how relational models are a collection of **integrated** records. By repeating columns in multiple tables, the tables become integrated.

Databases are also **self-defining**. Not only is the actual data stored in tables, but there are special tables that store information about the data tables. There is always

Table 10.1 CHEMICAL relation

Common name	Chemical name	Formula	Cost
Oxygen	Oxygen	O	$0.5 per liter
Hydrogen	Hydrogen	H	$0.25 per liter
Salt	Sodium	Na	$0.75 per kilogram
Carbon	Carbon	C	$0.50 per kilogram

Table 10.2 PRODUCT relation

Common name	Chemical name	Cost
Water	H2O	$1.00 per liter
Lye	Sodium Hydroxide (NaOH)	$2.00 per kilogram
Baking Soda	Sodium Bicarbonate (NaHCO3)	$0.50 per kilogram

Table10.3 INGREDIENT relation

Product name	Chemical name	Amount
Water	Hydrogen	2 1
Water	Oxygen	2 1
Lye	Sodium	1 1
Lye	Oxygen	1 1
Lye	Hydrogen	1 1
Baking soda	Sodium	1 1
Baking soda	Hydrogen	1 1
Baking soda	Oxygen	1 1
Baking soda	Carbon	3 1

Table 10.4 INSTRUCTION relation

Product name	Instruction
Water	Combine hydrogen and oxygen in a closed system and stir
Baking soda	Combine all ingredients with small heat and stir
Lye	Be sure to have protective gear! Combine all ingredients over medium heat and stir for 1 hour, let cool

at least one table that describes which columns are in which table and what data type each column is. The name of this table will vary (depending on the software) but will look similar to the following (Table 10.5).

Because the structure of the tables is stored in known tables, software can retrieve this information when needed and adapt to any changes in the database. Can you think of other information that a database software system might want to store about the data being stored?

While storing, retrieving, and describing data is the primary function of a database management system (DBMS), this type of software often does much

Table 10.5 Generic database description table

Table	Column	Datatype
CHEMICAL	CH_COMMON	CHAR(32)
CHEMICAL	CH_CHEMICAL	CHAR(32)
CHEMICAL	CH_FORMULA	CHAR(64)
CHEMICAL	CH_COST	FLOAT
PRODUCT	P_COMMON	CHAR(32)
PRODUCT	P_CHEMICAL	CHAR(32)
PRODUCT	P_COST	CHAR(30)
INGREDIENT	I_P_NAME	CHAR(32)
INGREDIENT	I_C_NAME	CHAR(32)
INGREDIENT	I_AMOUNT	CHAR(32)
INSTRUCTION	IS_P_NAME	CHAR(32)
INSTRUCTION	IS_INSTRUCTION	CHAR(512)

more because of the large volumes of data being stored. When a very large quantity of data is being stored (and possibly accessed and updated by many different people at the same time), the DBMS must provide very efficient and safe access to the data. There are different ways to store the actual data that can slow down or speed up the query. There are also different ways to order the query that may cause different speeds. In addition, coordinating different accesses to the data so that different users will not cause problems for each other is a very important aspect of databases. However, in this module most of this is transparent to the user and not what you are looking at. You will be concerned about speed issues (that might be under our control) and how to actually access the data.

One way that DBMS's can help speed access to the data is by using indexes. An index is used to help find information in a table. Instead of having to search a large number of rows (typically over a million rows), an index can list the location of the rows that have a particular piece of data (example: the rows in the enroll table for each student). This changes a sequential search (O(n)) into a direct access (O(1)) (see Module 9 for explanation and implication of Big-Oh notation). If you are doing a number of large searches, this difference can become quite significant. This is just for looking up rows in one table. When combining tables, the work becomes much larger. The combining tables is called a *join* which can often involve matching every row in one table with every row in another table (and then removing the rows that are not important). If the first table had 5 rows and the second table had 8 rows, the join of the two tables could have as many as 40 rows (5 * 8). This is an $O(n^2)$ operation! Joining three tables would be an $O(n^3)$ and joining 5 tables would be an $O(n^5)$ operation!

Answer It!

Now take the example database to a chemical plant

- 50,000 different ingredients used as chemicals
- 40,000 different products being produced

- each product using an average of 12.3 different chemicals and 10 different steps
- There needs to be an additional column in the instruction to indicate the order of the instructions to follow.

Q10.16: Assume that you need to list the chemical name of all the ingredients in 20 different products. About how many rows will be in the ingredient table? Assume that the table is not sorted and you don't have an index. How many comparisons will be needed to find all of the chemicals for all of the products? What if there is a direct index for the chemicals into this table? How did you get your answer?

Q10.17: What if you needed the chemical formula instead of the chemical name. What is the maximum number of rows that would result in the join? How many comparisons are needed now, if the result is not sorted and does not have an index, and if there is a direct index? How did you get your answer?

Q10.18: What if you not only needed the product name but also the product chemical name? What is the maximum number of rows that would result in the join? How many comparisons are needed now, if the result is not sorted without an index, and if there is a direct index? How did you get your answer?

(The actual time difference for the two different methods would even be worse because you did not take into account the time it takes to actually compute the join.)

10.7 Self-Defining Type 3 Part 2: Data Warehouses

One thing common to many databases associated with the sciences is that the data is typically not very dynamic. Data is not usually changing every second, and most updates (if they do happen in real time) are just adding new data (not changing or deleting existing data). When this is known about the data, things can be done to help speed access. If things are constantly changing, building indexes and pre-joining tables can actually slow down a database instead of speeding it up. But when data doesn't change often, some of this work can be pre-computed, allowing for significant improvements in speed.

Often this non-changing data (with pre-computed joins and other data) is stored in something called a *data warehouse*. The purpose of a data warehouse is to do many computations once (early) so that they need not be repeated and building many indexes to allow for quick access. However, data warehouses are rarely updated and often taken off-line to do a large update at once. Many of the databases (with large amounts of data) that a scientist might interact with might be stored in a warehouse for these reasons.

Discuss It!

Find two examples of large databases (online) that are associated with your favorite field of science. Can you tell if they are stored in a data warehouse? (You might not be able to tell this). Can you find out what tables actually exist?

Often, users will not directly access the database but will instead use a second piece of software to aid in finding the data. There is a language (called SQL: Structured Query Language) that has been designed to allow users to directly access a database. It has three major parts: ***select***, ***from***, and ***where***.

Example 1: Working with One Table

```
select CH_FORMULA
from CHEMICAL
where CH_COMMON='Carbon'
```

The *select* portion tells which columns to return. The *from* portion tells what table(s) to use. The *where* portion is used to limit the rows.

When working with data from more than one table, you have multiple tables in the *from* portion and specify the columns to use for the join in the *where* portion.

Example 2: Working with Two Tables

```
select CH_COST
from CHEMICAL, INGREDIENT
where I_P_NAME='Baking Soda' and I_C_NAME=CH_COMMON
```

Sometimes, you include join conditions and tables that are not in the final result because they are needed to get from one table to the other table (e.g. from the professor's name to the course name).

Example 3: Working with Three Tables
In this example, you needed the INGREDIENTS table to connect the PRODUCT and CHEMICAL tables.

```
select P_COMMON
from PRODUCT, CHEMICAL, INGREDIENTS
where CH_FORMULA='Na' and CH_COMMOND=I_C_NAME
and I_P_NAME=P_COMMON
```

SQL also has the ability to sort the data and group it by common fields. This allows you to write statements to tell how many ingredients in each product (as an example). SQL is a very powerful language that can be mastered by everyday users with some work, but it is often not directly used by most users because there is software that can build the SQL statement for most users. This even includes most spreadsheets (including EXCEL). It can interface with almost any database software and provide a wizard to help the users create the SQL commands by describing what is wanted. However, sophisticated users may need to use SQL to access data in the way they want. This is not uncommon in some science areas!

Answer It!
Write simple SQL statements for each of the following. If you have not covered the exact syntax, make a good guess! You will most likely be right if you take an example that is similar and change it!

Q10.19: What are the products that use ingredients that cost more than $4.00?
Q10.20: What are the formula names for all the ingredients in baking soda?
Q10.21: What are the instructions for creating lye?

10.8 Self-Defining Type 3 Part 3: Other Database Types

While relational databases are the most common databases, there are other types that are used (and some of these are found in the sciences). These include:

- spatial databases – These are often found in GIS systems. They include additional data about where in space (2D or 3D) items can be found and can tell the user how close things are.
- object-oriented databases - These are new types of databases and are not often found in the sciences. They are structured around items called "objects" which is a very complicated term and beyond the depth of this module.

- temporal databases – These also include time stamps associated with each item in the database allowing the user to track changes over time and can be found in certain areas of science.
- biological databases – These are databases fine-tuned to the specific needs of a sub-discipline of biology

Many databases found on the Internet are often distributed databases (with the data being stored across a number of different DBMS's on different machines often in very different geographical locations). These have unique challenges of their own including:

- Efficient access (needing to send large amounts of data across the internet for joins of tables stored in different locations)
- Coordinating updates (making sure if the data is duplicated that all copies of the data are always consistent even if some of the databases go down)
- Coordinating the data itself (sometimes when combining two or more existing databases there are problems - > the same data might have different names, some portions of the data might be missing in different original databases, and the data might actually have different values between the different databases)

Answer It!

Q10.22: Find two examples of scientific databases that are not a relational model. Explain what type of database each is. This might not be easy to be sure that your answers are correct. That is okay. Make a good effort.

Q10.23: Why might it be important to know if the database you are accessing is a distributed database?

10.9 Related Modules

- Module 3: Data Types: Representation, Abstraction, Limitations. Data storage information is covered.
- Module 11: Searching. Searching large datasets.
- Module 14: Data Organization and Analysis. Application of databases, including spatial databases is explored.

Reference

Tschopp E et al (2015) A specimen-level phylogenetic analysis and taxonomic revision of Diplodocidae (Dinosauria, Sauropoda). PeerJ 3:e857. doi:10.7717/peerj.857

Searching 11

11.1 Objectives

After completing this module a student should be able to:

- Describe a heuristic search
- Define definitions listed in the definitions section
- Implement a Basic Local Alignment Search Tool (BLAST) algorithm.
- Appreciate the limitations of manual execution of algorithms for analysis of large amounts of data
- Appreciate the need for appropriate software tools to conduct complex/large analysis
- Appreciate reasons for using a fast but less sensitive algorithm in some scenarios
- List two other uses of heuristic search techniques in scientific investigation

11.2 Definitions

- Algorithm
- Amino acid
- Heuristic search
- Polypeptides
- Protein
- Query sequence
- Sequence homology/percent identity
- Substitution matrix
- Target sequence

K. Brewer, C. Bareiss, *Concise Guide to Computing Foundations*,
DOI 10.1007/978-3-319-29954-9_11

11.3 Motivation

Once data is collected and stored in a computer system (typically in some sort of database), the scientist or engineer will need to find one or more data items that meet some sort of search criteria. For small data sets, we don't need to be concerned too much about the algorithm we use, nor its efficiency, as the entire process can happen fast and/or easily. However, when our data sets get large, or our search gets complicated (with many conditions, or even with a few "flexible" (or fuzzy) conditions) the algorithm we use becomes important to consider. This module will use a biological science example to examine some of the issues with searching.

11.4 Searching Amino Acids

Biologists have identified 20 different amino acids and each is designated with a different letter. Polypeptides are short strands (fewer than 50) of amino acids. Proteins are typically organized from long strands of amino acids. Assume you want to find a match between your "query sequence" that consists of 10 amino acids, against 10 different protein sequences of 99 amino acids each. Example sequences we will use throughout the first part of this module are given at the end of this module.

Answer It!

Q11.01: Calculate how many comparisons you will need to make for a single 99 amino acid protein sequence. (Remember a 10 amino acid sequence to 10 sequence comparison counts as "10" comparisons, not 1. For example, the first comparison for the first sequence would compare ALAPPPHDDN to KHPMLALTDQ – and that comparison would count as "10" since you would need to compare all 10 letters.)

Q11.02: How many comparisons for all 10 sequences?

Q11.03: If you had to compare your 10 amino acid sequence to a sequence that is more typical of a real protein (of 1000 amino acids), how many comparisons would be required?

Q11.04: Since humans make at least 50,000 different proteins, and if each was made from a 1000 amino acid sequence, how many comparisons would be required?

Q11.05: Now do the actual comparison with only the first protein sequence given, against the Query Sequence. Record and report any exact matches you find. Indicate what was the starting letter position. (1st, 2nd, 4rd, etc.).

Q11.06: It is likely you did not find an exact match in the previous question, but biologists have determined that sometimes nature allows substitution of one amino acid for another in the protein sequence, with no significant biological effect. Since we don't need to find an exact match, we should now look for close matches. So instead of exactly matching all 10 amino acids, we need to look for matches of only 9. List all the amino acid sequences you will need to look for if only matching nine. Use an "x" to indicate omitted letters. The search sequence omitting the first letter is already given for you. How many total search sequences?

xLAPPPHDDN

Q11.07: Determine how many sequences would need to be searched for if only matching eight. (Don't list them, just figure out the total number.)

Q11.08: The mathematical description of the above is "combinations". Use WolframAlpha to validate your previous answer. Type: combination [{1,2,3,4,5,6,7,8,9,10},{8}] into WolframAlpha. Was your original answer correct? What is the total number of combinations for matching 8?

Q11.09: Considering your answers to the previous questions, how many total sequences would be needed to be searched for, for all possible sequences from all 10 down to only three (e.g., xxxPPPxxxx or xLxPPxHDDx, etc.)? Use WolframAlpha.

Discuss It!

How many comparisons would need to be made for 10 down to only three exact matches for our Query Sequence for 50,000 different proteins, if each was made from a 1000 amino acid sequence?

Biologists have determined that some amino acid substitutions are more likely than others based on studying lots of sequences. This construct is called a substitution matrix. The substitution matrix allows us to score sequences for similarity even if they contain different amino acids (because you know that amino acids in the same group will act similarly).

Think about this concept in everyday use. You plan to make a cake and you have a recipe. The recipe calls for butter, but you are out of butter. Is there something else you can use? You need a substitution matrix to help us out.

Answer It!

Q11.10: Come up with some similarity scores for the listed items. You can use scores ranging from **−5** (poor match) to **+5** (best match) (Chart 11.1).

Chart 11.1 Similarity scores

	Oil	Butter	Sugar	Bear fat	Salt	Flour	Corn meal	Eggs
Oil								
Butter								
Sugar								
Bear fat								
Salt								
Flour								
Corn meal								
Eggs								

	A	C	D	E	F	G	H	I	K	L	M	N	P	Q	R	S	T	V	W	Y
A (Alanine)	**4**	0	-2	-1	-2	0	-2	-1	-1	-1	-1	-1	-1	-1	-1	1	-1	-2	-3	-2
C (Cysteine)	0	**9**	-3	-4	-2	-3	-3	-1	-3	-1	-1	-3	-3	-3	-3	-1	-1	-1	-2	-2
D (Aspartic Acid)	-2	-3	**6**	2	-3	-1	-1	-3	-1	-4	-3	1	-1	0	-2	0	1	-3	-4	-3
E (Glutamic Acid)	-1	-4	2	**5**	-3	-2	0	-3	1	-3	-2	0	-1	2	0	0	0	-3	-3	-2
F (Phenylalanine)	-2	-2	-3	-3	**6**	-3	-1	0	-3	0	0	-3	-4	-3	-3	-2	-2	-1	1	3
G (Glycine)	0	-3	-1	-2	-3	**6**	-2	-4	-2	-4	-3	-2	-2	-2	-2	0	1	0	-2	-3
H (Histidine)	-2	-3	1	0	-1	-2	**8**	-3	-1	-3	-2	1	-2	0	0	-1	0	-2	-2	2
I (Isoleucine)	-1	-1	-3	-3	0	-4	-3	**4**	-3	2	1	-3	-3	-3	-3	-2	-2	1	-3	-1
K (Lysine)	-1	-3	-1	1	-3	-2	-1	-3	**5**	-2	-1	0	-1	1	2	0	0	-3	-3	-2
L (Leucine)	-1	-1	-4	-3	0	-4	-3	2	-2	**4**	2	-3	-3	-2	-2	-2	-2	3	-2	-1
M (Methionine)	-1	-1	-3	-2	0	-3	-2	1	-1	2	**5**	-2	-2	0	-1	-1	-1	-2	-1	-1
N (Asparagine)	-2	-3	1	0	-3	0	-1	-3	0	-3	-2	**6**	-2	0	0	1	0	-3	-4	-2
P (Proline)	-1	-3	-1	-1	-4	-2	-2	-3	-1	-3	-2	-1	**7**	-1	-2	-1	1	-2	-4	-3
Q (Glutamine)	-1	-3	0	2	-3	-2	0	-3	1	-2	0	0	-1	**5**	1	0	0	-2	-2	-1
R (Arginine)	-1	-3	-2	0	-3	-2	0	-3	2	-2	-1	0	-2	1	**5**	-1	-1	-3	-3	-2
S (Serine)	1	-1	0	0	-2	0	-1	-2	0	-2	-1	1	-1	0	-1	**4**	1	-2	-3	-2
T (Threonine)	-1	-1	1	0	-2	1	0	-2	0	-2	-1	0	1	0	-1	1	**4**	-2	-3	-2
V (Valine)	0	-1	-3	-2	-1	-3	-3	3	-2	1	1	-3	-2	-2	-3	-2	-2	**4**	-3	-1
W (Tryptophan)	-3	-2	-4	-3	1	-2	-2	-3	-3	-2	-1	-4	-4	-2	-3	-3	-3	-3	**11**	2
Y (Tyrosine)	-2	-2	-3	-2	3	-3	2	-1	-2	-1	-1	-2	-3	-1	-2	-2	-2	-1	2	**7**

Fig. 11.1 BLOSUM62 amino acid substitution index

So instead of a search for matches, we will need to have an efficient algorithm to compute and compare a similarity score. The current popular (and effective) algorithm for amino acids is called BLAST (Basic Local Alignment Search Tool) and one family of matrices for doing the scoring is called BLOSUM (BLOcks of Amino Acid SUbstitution Matrix). This set of matrices has been empirically determined from comparing and aligning proteins with same function but different structure. A commonly used, general-purpose matrix in this group is BLOSUM62, shown here (Fig. 11.1).

In this table, each of the 20 amino acids is represented by a capital letter. To compare the similarity of two amino acids, find one on the left-hand column and the other in the top row; the intersection shows the similarity score for the two amino acids.

11.5 BLASTP Algorithm

The BLASTP (Basic Local Alignment Search Tool for Protein to Protein searches) Algorithm is widely used for searching protein sequences due to its speed (efficiency). The concept of BLASTP is to find and rank possible match locations – but without guarantee that the best location will be found. This "without guarantee" (the guarantee is sacrificed for speed) makes this a heuristic algorithm. Possible matches (called alignments) are initial found using smaller "seed words" (used for speed). Then the total quality of each initial alignment is calculated and the top alignments are eventually reported.

Discuss It!
How might the BLASTP algorithm (with some modification) be useful for searching large datasets of astronomical data? Meteorological data? Paleontological data?

We will learn the basics of the algorithm by doing a simple by-hand search based on our previous example sequences.

Step 1: Break up the Query Sequence into smaller groups called Words.
The user chooses the size, k, of the Words. Then all possible Words are determined. In our example with a Query Sequence of ALAPPPHDDN, if k = 3, we would create and use the following "Words":

`ALA, LAP, APP, PPP, PPH, PHD, HDD, DDN`

Answer It!
Q11.11: Determine all the k = 3 Words for the following Query Sequence: `PQGEFG`

Step 2: For each Word, find all Neighborhood Words above a Threshold score.
Using the selected substitution matrix (like BLOSUM62) a score can be calculated for each Word, and for all possible combinations based on substitutions. For example, if the Word is `ALA`, the score would be 4 + 4 + 4 = 12. For a Neighborhood Word with a `V` substituted for the `L` (i.e., `AVA`), the score would be 4 + 1 + 4 = 9. All possible Neighborhood Words are determined and scored. (There are 1540 of them for a 3-letter Word!) Save all the Neighborhood Words that are above or equal to `T`, a Threshold value. Repeat for each Word.

Answer It!

Q11.12: Determine the score for the following Neighborhood Words to `PPP`:

`PAP, PLP, PRP, PNP, PDP, PCP, PQP, PEP, PGP, PHP, PIP, PLP, PKP, PMP, PFP, PSP, PTP, PWP, PYP, PVP`

Step 3: Find all exact matches in the Sequences (database) for each Neighborhood Word. This step can be rapidly accomplished with an efficient search tree.

Answer It!

Q11.13: Find the starting positions for the Neighborhood Words in Q11.12 that had a score greater than or equal to a Threshold value of 13, anywhere in any of the 10 protein sequences in our example.

Step 4: Extend the matches from the Neighborhood Words to High-Scoring Segment Pairs (HSPs)

This step consists of a number of sub-steps. First, the Query Sequence is aligned with the Sequence around the Neighborhood Word. Let's suppose we use the Neighborhood Word: `PEP` (from the original query Word `PPP`), and we find and then align

```
Query Sequence:                A L A P P P H D D N
Database Sequence:  ... C F I D M L L P E P Y E W N G K M P ...
```

Next, extending out from the Word, determine the scores for each pair. For our example, we would have:

```
Query Sequence:               A L A P P P H D D N
Score:                       -1 4-1 7-1 7 2 2-4 6
Database Sequence:  ... C F I D M L L P E P Y E W N G K M P ...
```

Finally, determine the length of the maximum set of pairs where the positive value substitutions outweigh the negative substitutions (called mismatches or gaps). Designate the maximum set as a HSP. In our example, we would have `PEPYE` as a HSP:

```
Query Sequence:               A L A P P P H D D N
Score:                       -1 4-1 7-1 7 2 2-4 6
Database Sequence:  ... C F I D M L L P E P Y E W N G K M P ...
```

In some cases, it is desirable (from a practicability standpoint) to set a maximum length of an HSP.

Answer It!

Q11.14: For the match of our Neighborhood word as your answer to Q11.13, determine the HSP.

Step 5: Reduce the list of HSPs based on a Cutoff Score (S) and Expectation Value (E)

Eliminate all HSPs in the list that are below a Cutoff Score, S. The S value is determined from the statistics of all the HSP scores, to ensure the results are statistically significant. The E value is also determined from your HSPs and represents the number of HSPs that could be expected by chance to be above the S-value.

Step 6: Report the HSPs.

The remaining HSPs are presented, in decreasing order of scores, showing the match. The following is a more complex example than what we have been using, but represents actual output from the BLASTP program:

```
Query: 207 VDGQRRLIAVVMGADSAKGREEEARKLLRWGQQNFTTVQILHRGKKV 253
           VD  RR  ++    D    R +E  + L+WG Q F  ++I+  G  +
Sqnce: 130 VDFNRRADSLQKNQDLEFERNKERFEFLKWGSQAFCNMRIIPPGSGI 176
```

The 207 and 253 refer to the positions in the Query Sequence that is being shown (same for the 130 and 176). In the line between the Query and Sequence, Letters indicate exact matches, + indicate substitutions that have a positive value, and blanks indicate substitutions that are negative.

Discuss It!

The MEGA (Molecular Evolutionary Genetics Analysis) program implements the BLAST algorithm to search a large database of genetic (DNA and protein sequence) information from a wide variety of species.

11.6 Computing Questions

- For the different integers in your module, what data size(s) do you think were used to represent them? Why?
- For the different floating point numbers in your module, what data size(s) do you think were used to represent them? Why?

- What are the different sources of errors that might be encountered in the computational work done in this module?
- For the largest typical size of data used in the area of science studied in this module, how much space is needed? Why?
- Explain one algorithm used by the computer while working on your module.
- Was compression used in your module? What type of compression: loseless or lossy? Explain how you came to your conclusion.

11.7 Related Modules

- Module 10: Self-Defining Data: Compression, XML, and Databases. Databases are.
- Module 14: Data Organization and Analysis. Databases are explored and explained.

Acknowledgement The original idea for this module was by Dr. Greg Long

Ten Protein Sequences of 99 Amino Acids

```
MMAMWWCDNWFGADLHYSLHDIEDYVNMLSTEPCGMIVWYPPPHDDNTVRMAAVGNPHCYFHASG
NLGNAYNWHWYCGCCDLLCQY NPKCGWLKKIFAI

KHPMLALTDQNMHRTIWCERHIGRDQEYNYIIANRKCVSNDMEYREIWKAGWPGKFMGFRTTQLA
CMLICKKAILYQRQHCTTNKQNKGNVSHQPVECE

CNFWTNPNRNIPEDHALFNA GHFDCPPTAPHINGTLLNDQHFWSKEGRQAEVFGENNKVLNMQAM
SFIVHVKCQCWHWFFRGMKDIR NSSPKCNLVSRW

FCMKDLPCRPSIRAEDSGFKPCVCFIDMNLPEPYPRYGKMPNYYTYISCTFSCWEHSVNGHDWNT
TFLHYMNMLKSN HYTSRSESYKWKDRTVSTMFIC

DINHRCQLGSMQMSHPDEVVCVCLNFGESMYPHPLCQDYAYCQCMMSKTLCQIPAFKCQHVHKCT
DQTGVYI GGKCDQAPYVGNMTAENYILHTIREPQ

RMAYVDMHYWHWLFLTQWNPEQIPCHNKHILDTCHSINPPYENMLTNACERGEFVGQMPQVVYKN
WNDILSKQRLWCFYWR PHGDAMMAWQDMEQLFVE

FETEEMYICWMRVSMSLLQWANYYLKESNHYLCWSVHPWLYGFAMRVFDTPVFAYFVVYKEITAK
PIVTEPYFDMVCAHAATSFPVCS IRPFNDFGSVY

ALICNLRGQAIHADDHGDFWLCKYLWGCPEKALLFCFKMILATSGENLTKQRLEPVSLCMLYVST
MDLVQPMSLCKCIPKVPVKDSPW WREMVPVMPTC
```

CCFIVFHIYHLYSPSVCKGYQWASNRVQWAKQPQLWNCHFWYQAPGMSQQCPSSNKTYMAPWLLY
YIYDRPAHNQWLQAHTPC MKWISQICVEEFVIIS

SEECMFLCAHGRLFHTMREHHKLCLMFWMHAGDLPHHFIASWFTENEFCNDMQVWNIVSCIDWLR
HVNYGQCGIDMKCCVLVPWNRWCMS IRRTIQGET

References

BLAST Algorithm on Wikipedia: https://en.wikipedia.org/wiki/BLAST

MEGA home page – www.megasoftware.net with information from Tamura K, Peterson D, Peterson N, Stecher G, Nei M, Kumar S (2011) Molecular biology and evolution 28:2731–2739

National Center for Biotechnology Information: www.ncbi.nlm.nih.gov

12 Curve Fitting

12.1 Objectives

After completing this module, a student should be able to:

- Find an equation to approximate a set of data points
- Determine the error of such an equation
- Define splines
- Describe limitations of curve fitting methods

12.2 Definitions

- Curve fitting
- Splines
- Sum of the squares
- Regression

12.3 Motivation

After collecting data, scientists and engineers often want to determine which mathematical equation, if any, that best matches their data. Best is defined as when the calculated error (calculated in some fashion) between the model (the result from the mathematical equation) and the known data is minimized. This is commonly referred as "curve fitting" which is the subject of this module.

For the following tasks, you will be using both of these sets of (x, y) data:

Set 1: (1, −5), (2, −12.4), (3, −15.7), (4, −15.1), (5, −10.5), (6, −1.9), (7, 10.7), and (8, 27.4)

K. Brewer, C. Bareiss, *Concise Guide to Computing Foundations*,
DOI 10.1007/978-3-319-29954-9_12

Set 2: (0, 1.2), (1.5, 2.5), (2.5, 4.5), (4.0, 5.3), (6.5, 4.5), (8.1, 2.3), (9.3, 0.7), (11.3, −2.0), (13.0, −3.9), (15.5, −4.2), (17.5, −2.6), and (19.0, −0.5)

12.4 Fitting "By Hand"

To gain a better understanding of the underlying principles behind curve fitting, we will perform the basic steps "by hand" and without the "aid" of a graph. We will, however, use a computer spreadsheet to make the mechanics of the calculations less tedious.

Step 1: Enter the data into a spreadsheet (with x values going in column A and the corresponding y values going in column B).

Step 2: Column C will represent a linear equation, Column D will represent a quadratic equation, Column E will be cubic, and Column F will be a polynomial of degree 4. We will use the constants ***a*** through ***e*** as applicable in each of our equations to calculate the ***y*** value given an ***x*** value.

Step 3: Construct each equation to calculate ***y*** using additional cells to represent the constants so that you may modify the constants later. For now, make all the constants equal to 1. (See Fig. 12.1).

Step 4: In other columns to the right, calculate the differences between the dataset (actual) y and the y values computed by each function. Also calculate the squares of the differences

Discuss It!

The difference between the actual y and the computed y is considered the error. But when considering the overall magnitude of error, we don't care about the direction (greater or less than) of the error. How might we remove the "direction" mathematically? Think about your answer in the context of the step above

	A	B	C	D	E	F	G	H	I	J	K	L	M	N	O
1			Linear	Quadratic	Cubic	Poly-deg 4									
2			y=bx+a	y=cx^2+bx+a	y=dx^3+cx^2+bx+a	y=ex^4+dx^3+cx^2+bx+a									
3		a	-30	-3	-100	-10									
4		b	2	-2	-5	-2									
5		c		1	-3	3									
6		d			1	-4									
7		e				1		Linear		Quadratic		Cubic		Poly-deg 4	
8							Sum	139	3393	-137	3013	267	71071	-4077	6422273
9	X	Y	y	y	y	y		Y-y	(Y-y)^2	Y-y	(Y-y)^2	Y-y	(Y-y)^2	Y-y	(Y-y)^2
10	1	-4	-28	-4	-107	-12		24	576	0	0	103	10609	8	64
11	2	-10	-26	-3	-114	-18		16	256	-7	49	104	10816	8	64
12	3	-30	-24	0	-115	-16		-6	36	-30	900	85	7225	-14	196
13	4	-10	-22	5	-104	30		12	144	-15	225	94	8836	-40	1600
14	5	-5	-20	12	-75	180		15	225	-17	289	70	4900	-185	34225
15	6	0	-18	21	-22	518		18	324	-21	441	22	484	-518	268324
16	7	10	-16	32	61	1152		26	676	-22	484	-51	2601	-1142	1304164
17	8	20	-14	45	180	2214		34	1156	-25	625	-160	25600	-2194	4813636

Fig. 12.1 Example spreadsheet result (Your spreadsheet may look similar to this one (but with correct data))

Step 5: For each column from the previous step, calculate the sum.
Step 6: Modify your constants (keep them as integers) until you find the best fit (represented by the lowest sum of squared errors) for each equation (linear, quadratic, etc.).

Answer It!

Q12.01: For Set 1, report your "best fit" function (either linear, quadratic, etc.) you determined (write your equation with your determined constant values), how off were you (sum of the squared values), and approximately how long did it take to figure this out (minutes).
Q12.01: For Set 2, report your "best fit" function you determined, how off were you, and approximately how long did it take to figure this out.

12.5 Fitting By Hand with Graphing Aid

Not only can computing make the mechanics of curve fitting calculations less tedious, we can use computer visualization to enhance the process.

Step 1: Launch http://www.shodor.org/interactivate/activities/DataFlyer/ and enter the data (correctly) in the data grid and plot it. (The program expects each x-y pair of data to be on separate lines, separated by a comma.) Make sure you adjust the view of the graph to see all your data points.
Step 2: In the function, f(x) =,box, type in a function (with initial constant values of 1) and make sure the "Exponents Change" box is NOT checked. For a quadratic function, you would type the equation as shown in Fig. 12.2.
Step 3: After entering the equation and hitting enter, you should see your constants change to colored values that correspond to the color sliders (Fig. 12.3).
Step 4: You can now use the sliders to modify the function's constants to try to make the equation (model) fit the data. You can adjust your slider limits if necessary – but keep the "step"s equal to one.
Step 5: Change the constants until you find the best fit.
Step 6: Once you are happy with your results, find on the page where it shows you the sum of the squares of the differences (or deviations).
Step 7: Click the "Show Squares" checkbox on and off to see a visual representation for your "sum of squares" values.

Fig. 12.2 Quadratic function entry

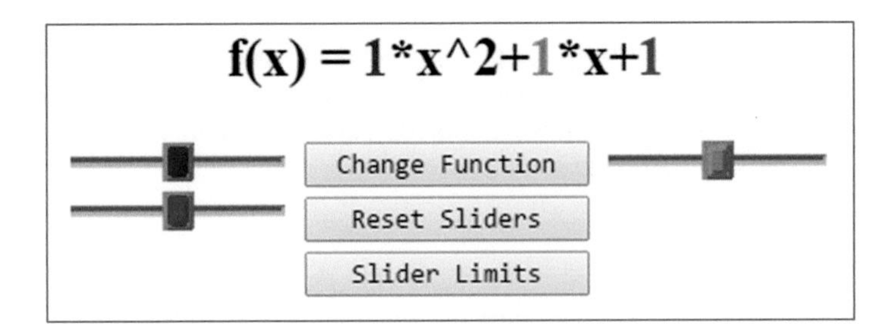

Fig. 12.3 Interface result

Answer It!

Q12.03: For Set 1, report your "best fit" function you determined (write your equation with your determined constant values), how off you were (the sum of the squares of deviations), and approximately how long it took you to figure this out (minutes).

Q12.04: For Set 2, report your "best fit" function you determined, how off you were, and approximately how long it took you to figure this out.

12.6 Fitting via Numerical Analysis (Regression)

Now that we have explored the inner workings of fitting, and the trial-and-error nature of achieving the "best fit", it is time to use common algorithms to quickly and accurately find the best curve fit. We will use WolframAlpha which has built-in functions to find "linear fit { }", "quadratic fit { }", and "cubic fit { }" – where your dataset is entered between the curly braces.

Answer It!

Q12.05: For Set 1, report your linear "best fit" function you determined (write your equation with your determined constant values), how off were you (coefficient of determination), and approximately how long did it take to figure this out (minutes).

Q12.06: For Set 2, report your linear "best fit" function you determined, how off were you, and approximately how long did it take to figure this out.

Q12.07: For Set 1, report your quadratic "best fit" function you determined (write your equation with your determined constant values), how off were you (coefficient of determination), and approximately how long did it take to figure this out (minutes).

Q12.08: For Set 2, report your quadratic "best fit" function you determined, how off were you, and approximately how long did it take to figure this out.

Q12.09: For Set 1, report your cubic "best fit" function you determined, how off were you (coefficient of determination), and approximately how long did it take to figure this out.

Q12.10: For Set 2, report your cubic "best fit" function you determined, how off were you, and approximately how long did it take to figure this out.

While the algorithm that Wolfram Alpha uses to do regression is not known, published regression algorithms are well known and easily implemented in computing. To obtain the best curve fit, a minimization of errors is determined. Errors are determined via squaring the vertical offsets between the curve/line and each data point. Squares are used so that the magnitude of the offset and not the direction of the offset is considered.

The best curve fit for a linear curve is called a linear least-squares line, naturally, with linear variables slope and intercept (b & a, respectively). The minimization of error results in a system of linear equations. These two linear equations (considered as a system of linear equations) can be written in matrix form. Matrix equations can be created to directly compute a and b, which can be easily programmed in a computer.

12.7 Fitting via Excel

For these two-variable datasets, we can use built-in functions and Solver in Excel to minimize the sum of squares of the differences while changing the equation parameters. Recall from Module 7 how the Solver tool can be used to find the best parameters. For curve fitting, we will solve for the best constants in our model equation that minimizes the sum of squares. Modify your previous spreadsheet to use the solver tool to calculate the best constant values.

Answer It!

Q12.11: For Set 1, report the linear "best fit" function you determined (write your equation with your determined constant values), how off were you (sum of squares), and approximately how long did it take to figure this out (minutes).

Q12.12: Repeat the above question for Set 2.

Q12.13: For Set 1, report the quadratic "best fit" function you determined (write your equation with your determined constant values), how off were you (sum of squares), and approximately how long did it take to figure this out (minutes).

Q12.14: Repeat the above question for Set 2.

Q12.15: For Set 1, report the cubic "best fit" function you determined (write your equation with your determined constant values), how off were you (sum of squares), and approximately how long did it take to figure this out (minutes).

Q12.16: Repeat the above question for Set 2.

Discuss It!

Where might computational errors enter the process? What do you think the Big-O of your algorithm is and why?

Discuss It!

Often, a real life datasets do not involve one equation but multiple equations. For example, a seismograph trace usually is composed of three processes and therefore could be represented by least three equations: one for the pre-earthquake event, one for the actual earthquake, and one or more for the aftershocks. One technique used to represent multiple equations on one line is called "splines". How might splines handle the appropriate "connection" between equations?

12.8 Computing Questions

- For the different integers in your module, what data size(s) do you think were used to represent them? Why?
- For the different floating point numbers in your module, what data size(s) do you think were used to represent them? Why?
- What are the different sources of errors that might be encountered in the computational work done in your module?
- Explain one algorithm used by the computer while working on your module.

12.9 Related Modules

- Module 3: Data Types: Representation, Abstraction, Limitations. Computational limits/errors (e.g. round-off, overflow, underflow, limited precision, non-reproducible computations) are examined further.
- Module 5: Procedures: Algorithms and Abstraction. Introduces concepts of algorithms and their performance when considering numerical methods.
- Module 7: Iterative Solutions. Uses goal seek and solver to find solutions to equations.
- Module 8: Solving Sets of Equations. Extends ideas for solutions of equations to simultaneous systems of equations.

13 Optimization

13.1 Objectives

After completing this module, a student should be able to:

- Describe the components of an optimization problem.
- Define Objective Function.
- Appreciate the Simulated Annealing and Genetic Algorithm techniques.
- Construct an Objective Function with parameters and constraints for a system that needs the best solution known.

13.2 Definitions

- Constraints
- Genetic algorithm
- Objective function
- Optimization
- Parameter
- Simulated annealing

13.3 Motivation

Optimization is often phrased as *the process of finding the best solution.* There are numerous occasions in science and engineering (some examples are included in this text in other modules) where finding the best solution is needed. If the problem is straightforward, finding the best can be achieved through execution of an algorithm (least squares curve fitting is a common example). In Module 12, you essentially performed an optimization when you found the best fitting line for a set of data points. You minimized the error between the line solution for y's, and the data y's.

K. Brewer, C. Bareiss, *Concise Guide to Computing Foundations*,
DOI 10.1007/978-3-319-29954-9_13

However, if the problem is complex or has many independent variables/parameters, finding the "best" may be difficult or computationally too expensive. For many real-world science and engineering problems, optimization techniques using computational algorithms are needed. This module will introduce you to a few optimization algorithms that can be applied in many scientific disciplines.

13.4 What Makes Up an Optimization Problem?

Any optimization problem has the following components:

- A set of independent variables or parameters that one wishes to find (find the *best* ones).
- A set of constraints on the variables or parameters.
- Defined relationship(s) or equation(s) between the variables that defines the problem, and
- A definition of what is *best*.

For the linear curve-fitting problem, the relationship was defined as a linear equation: $y = mx + b$. The *best* was a minimization of error (squared differences between the linear's equation's calculated y's and the data y's). The independent variables were the slope and intercept (m and b). And the constraints were that m and b had to be real numbers.

Discuss It!
How would you frame finding the roots of an equation as an optimization problem?

13.5 What Is the "Language" of an Optimization Problem?

Often, as can be expected, optimization problems are framed mathematically, so the "relationship" is formally called the Objective Function, O(x), and the *best solution* is defined as a maximization or minimization of that function.

Answer It!

Q13.01: State what the variable(s), constraints, Objective Function, and whether it is a minimization or maximization problem for the following:
"Find the roots of the following equation $y = x^2 + 5x + 2$ for all $x < 0$."

Most scientific optimization problems, however, aren't initially mathematical. They are problems that involve various data and relationships, and finding the *best* for that set of data and relationships is the goal. Nevertheless to solve these problems, scientists and engineers will first frame their problem mathematically.

13.6 Working Through the Setup of an Optimization Problem

To explore all this, think about a non-mathematical optimization problem that you have already solved.

Discuss It!
When you enter a classroom and choose a seat (or when you seat yourself at a restaurant), why are you sitting where you are and not at some other location in the room?

In order for you to make a decision where to sit, you likely went through an optimization process in your mind. [Note: If you are a typical student, you have likely chosen to sit in the same location during each class period. So for this exploration, you want to think about your thought process and decision where to sit during that "first day."]

Answer It!

Q13.02: List all the reasons that went through your mind that you considered when choosing where to sit, when you walked into the classroom (on the first day). Your reasons may be different from others, so also list other reasons that you think others might have used.

Each separate reason is a separate variable/parameter that you incorporated into your optimization relationship. You can quantify each of those parameters into numerical parameters (for example, *wanting to sit near an exit* is equivalent to a *distance from exit* parameter).

Answer It!

Q13.03: From each of your reasons, create a list of quantifiable variables/parameters. Also include the optimum direction and/or value for each parameter, the possible range of each parameter (for all seating locations), and define a variable letter. (e.g. if you are wanting to sit near an exit, the optimal direction is: minimum distance is optimal, the possible range is: 1 to 10 m, and a variable name could be: d).

You now need to construct the Objective Function from the list of variables. For this example, the Objective Function can be represented by a summation of each parameter, which will be computed for parameter values representing each possible sitting location in the room. But when constructing the summation, you need to ensure that the relative range of each term is appropriate. For this problem, what is "appropriate" is based on the relative importance of each reason/parameter.

For this seating example, say you only have two parameters: *brightness of the lighting,* b, and *distance from exit*, d. Let us assume brighter is better, and closer to exit is better with brightness measured in lumens and distance in measured in meters. Assume you want the *distance from exit* parameter to have twice as much importance as your *brightness of the lighting* parameter. Further assume that your measured brightness values at each desk can range from 500 to 1500 lumens, and your distances from 1 to 10 m. To create an Objective Function that represents your desire for twice as much importance on distance than brightness, you would start with each term ranging from 0 to 1 (with best equaling a value of 0):

$$\begin{array}{c} O() = (b - 500)/1000 + (-d + 10)/9 \\ \textit{brightness} \quad + \quad \textit{distance} \end{array} \tag{1}$$

To then account for the doubled importance of the distance parameter, the Objective Function would be:

$$O() = (b - 500)/1000 + 2*(-d + 10)/9 \tag{2}$$

which simplifies to:

$$O() = b/1000 - 0.5*d/9 \tag{3}$$

since constant terms in the equation are not important in determining parameter values to achieve a minimum function value.

Note that each term was linear – that is, the parameter relationship was represented by a straight line. Optimization techniques aren't necessarily restricted to linear problems, but they are easier to solve and should be desired over non-linear relationships when possible.

Answer It!

Q13.04: Determine and write your simplified Objective Function for your personal seating problem.

Q13.05: Using a problem similar to the Out on a Limb problem in Module 7, determine the parameters, Objective Function, constraints, and Goal (max or min). Assume you want to optimize the design to maximize the value of the sign (which equals benefits less costs). Assume the longer the tube, the more benefit ($100 per m the sign is extended out from the building, since it will be more visible, with a minimum length of 0.001 m), and the cost of the materials is based on the volume of the metal used in the tube ($0.10 per cubic cm). The minimum tube wall thickness is 1 mm.

13.7 Solving an Optimization Problem

Once you have your objective function defined (and simplified), the goal (minimize or maximize), the parameters (and ranges), and any problem constraints, you can now find the best value. For your seating problem, you could simply calculate the Objective function value for each seating location and then pick the lowest (or highest, depending how you set up your Objective Function) – which would tell you where you should sit!

But for most "real-world" optimization problems, it is not practical to calculate the Objective Function at each possible solution location due to a high number of possible solutions (recall the complexity discussion in Module 9). So you need a more efficient way to either calculate the best or at least have a reasonable expectation of finding the best.

Practically, there is no one-best way to find the best solution. Various techniques have been developed to improve the performance of the process (or even be able to arrive at a solution). Selecting which optimization technique is still somewhat a "dark art" by scientists and engineers. Two methods will be introduced in the remaining part of this module, and then the Simplex Method for optimizing linear programming problems will be discussed last.

13.8 Simulated Annealing

The Simulated Annealing method is an algorithm that uses a thermodynamics analogy of slow-cooling metals to achieve purer crystal structures. This *annealing* allows atom redistribution that is increasingly restricted (lose mobility) during cooling. For the optimization analogy, in your solution space you will initially easily allow your current solution to travel to worse solutions, but will slowly reduce that flexibility as the "temperature" goes down.

The overall algorithm is:

1. Pick a solution, x, and evaluate the objective function, O(x)
2. Pick a "nearby" solution, x_i, and evaluate the objective function, $O(x_i)$.
3. Move to x_i if
   (i) $O(x_i)$ is better than O(x)
   (ii) $\exp(-1 * (\text{abs}(O(x_i) - O(x))/T) > \text{random}(1)$
4. Reduce the temperature, T, by factor r – $T_{i+1} = T_i * r$
5. Go back to step 2. End when temperature is sufficiently "cold" and improvement has stopped.
6. Keep track of best-ever solution and use that as your *optimum*.

Fig. 13.1 Simulated Annealing NetLogo model

The key to simulated annealing is in designing how to determine the "nearby" solution and in getting the temperature variable right.

Below are three NetLogo simulations available to help you understand how Simulated Annealing works and how it can be implemented.

- In the *Models Library*, search for *Simulated Annealing* (Stonedahl and Wilensky 2009). This is a basic model to solve a simple system (Fig. 13.1).
- Simulated_Annealing_2D.nlogo (available on the textbook website) is a simple 2D problem (two variables, x and y) that provides a visualization of the solution technique. The solution domain is given by the colored patches (the red one is the ultimate best), and the current solution is given by a green turtle. Running the simulation will show you how the turtle traverses the solution space (Fig. 13.2).
- Simulated_Annealing_for_Beam.nlogo (available on the textbook website) is an implementation of the algorithm for a problem similar to the *Out on a Limb* problem in Module 7. The visualization is a cross-section of the tube (out from the wall on the left). Intermediate solutions are continuously shown as the algorithm progresses, with the overall best solution shown in yellow at the end of the run (Fig. 13.3).

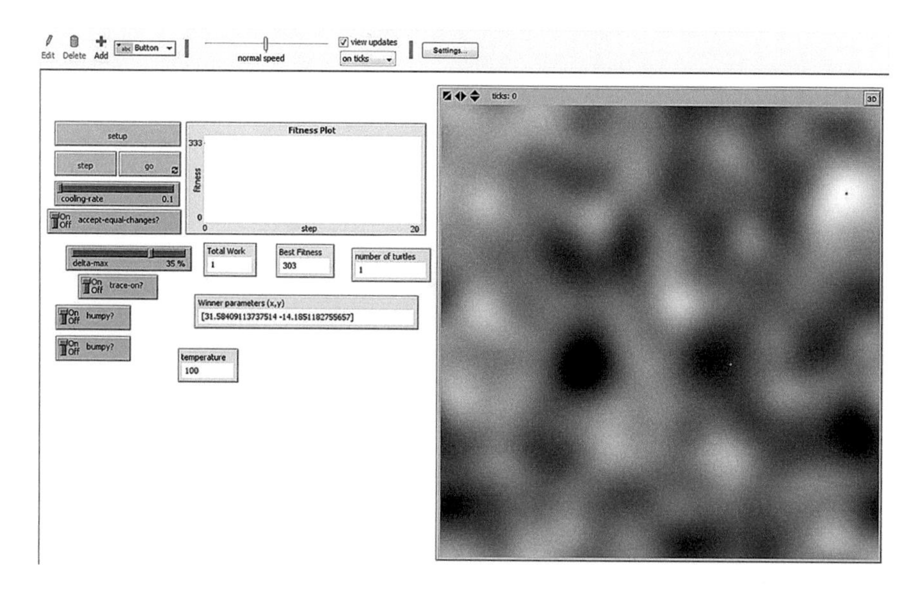

Fig. 13.2 Simulated_Annealing_2Dvis NetLogo model

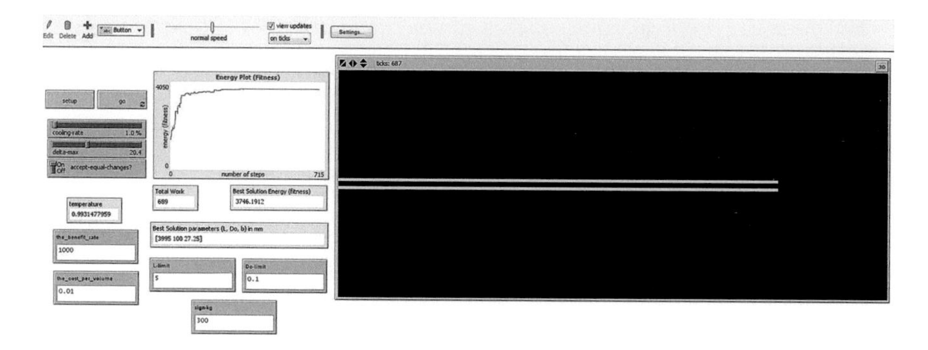

Fig. 13.3 Simulated_Annealing_for_Beam NetLogo model

Discuss It!

Run each model and explore the codes to understand each's implementation and any underlying limitations. In particular run the Simulated_Annealing_2D.nlogo model and see how the various "controls" affect the efficiency of reaching a solution (i.e., the amount of work), and how good the best-ever solution is compared to the ultimate best solution. Note that the solution algorithm has to overcome local peaks (i.e., go through valleys of worse solutions) to get to the ultimate best solution peak. Observe

(continued)

how the algorithm parameters (cooling-rate, r; largest possible move, delta-max; and allowing moves if no improvement, equal-changes) affect the ability of the algorithm to traverse the solution space and find the highest peak.

Answer It!

Q13.06: Using the *Simulated_Annealing_for_Beam.nlogo* simulation, determine the *optimum* configuration assuming the benefit rate is \$100 and the cost is \$0.01 per volume. Limit tube length to a maximum of 5 m. Make sure the sign mass is 45 kg and a safety factor of 2. Does the placing of limits on the tube length (L-limit, tube length in meters) and tube thickness (Do-limit, tube diameter in meters) affect your optimum?

Q13.07: How much work (which is equal to the number of solutions computed) was done to determine your solution? If you assume that your three parameters (L, Do, and b) can vary in increments of 1 mm, how many possible solutions are there?

Discuss It!

For the above problem, each solution can be calculated relatively fast and so evaluating large numbers of solutions doesn't take very long and the effort to implement an optimization algorithm may be greater than any savings (over just calculating each solution and picking the best). What about for more complex problems, where solution calculation may take hours for each one?

The simulated annealing algorithm isn't set in stone and variations have been proposed (and encouraged). One potentially beneficial modification is to periodically go back to the best-ever solution and continue from there (instead of picking a nearby solution).

Discuss It!

What might be a disadvantage of a modification that results in going back?

13.9 Genetic Algorithm

The Genetic Algorithm method mimics biological evolution in order to find the *best* solution. It uses the ideas of each calculated solution having a unique *gene sequence* (that is, a sequence of parameter values), with each calculated solution considered a member of a population. New generations of solutions are generated by reproduction of solutions based on selection of two *parents*, gene sequence cross-over

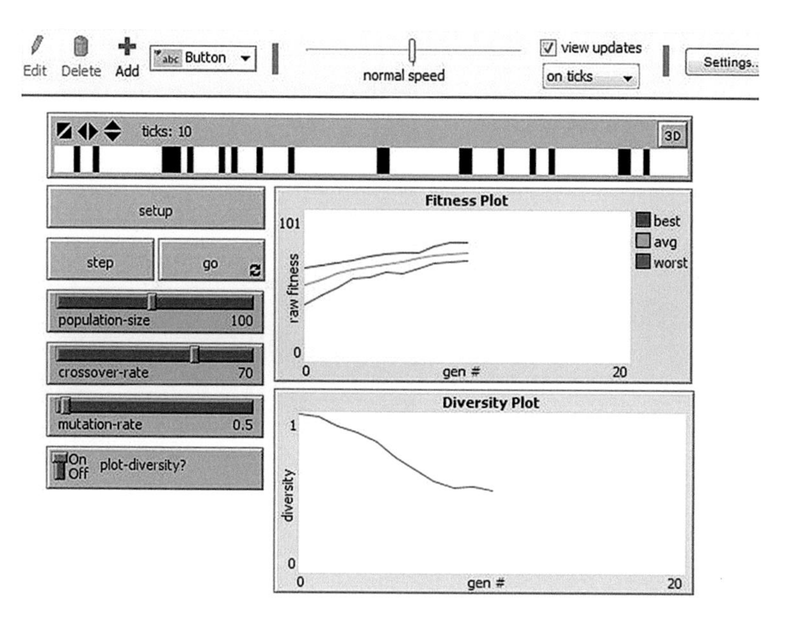

Fig. 13.4 Genetic Algorithm NetLogo model

between the two parent's genes, and random mutation of genes. Typically the *best* member of the current generation is *cloned* to continue on in the next generation. The resulting children make up the next generation as all the parents die off. New generations are created until little improvement in the best-ever solution is noted.

Below are three NetLogo simulations to help you understand how the Genetic Algorithm method works and can be implemented.

- In the *Models Library*, search for *Genetic Algorithm*. This is a basic model to solve a simple system (Fig. 13.4).
- GeneticAlgorithm_2D.nlogo (available on the textbook website) is a simple 2D problem (two variables, x and y) that provides a visualization of the solution technique. The solution domain is given by the colored patches (the red one is the ultimate best), and the current population of solutions is given by green turtles. Running the simulation will show you how the algorithm works through the solution space (Fig. 13.5).
- GeneticAlgorithm_for_Beam.nlogo (available on the textbook website) is an implementation of the algorithm for a problem similar to the *Out on a Limb* problem in Module 7. The visualization is a cross-section of the tube (out from the wall on the left).). Intermediate solutions are continuously shown as the algorithm progresses, with the overall best solution shown in yellow at the end of the run (Fig. 13.6).

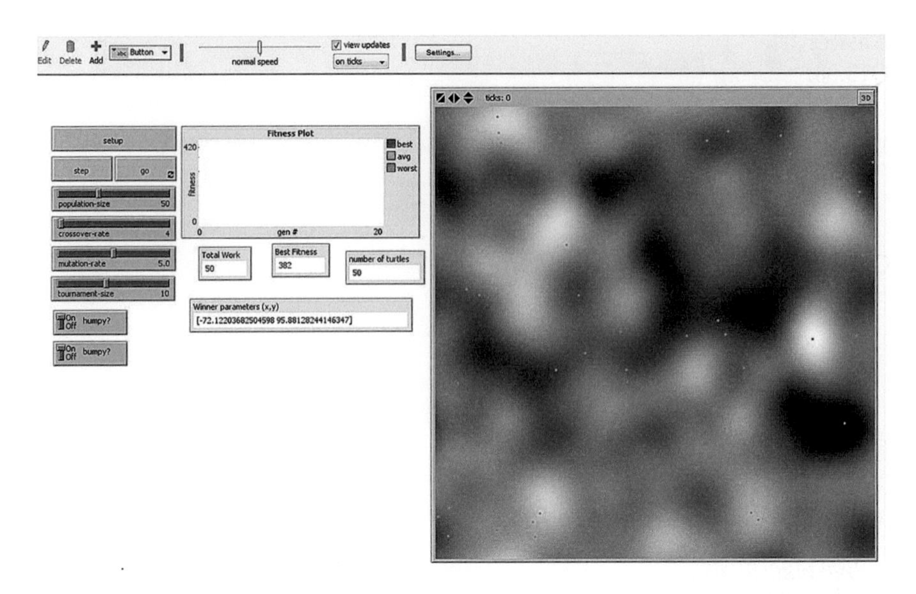

Fig. 13.5 GeneticAlgorithm_2Dvis NetLogo model

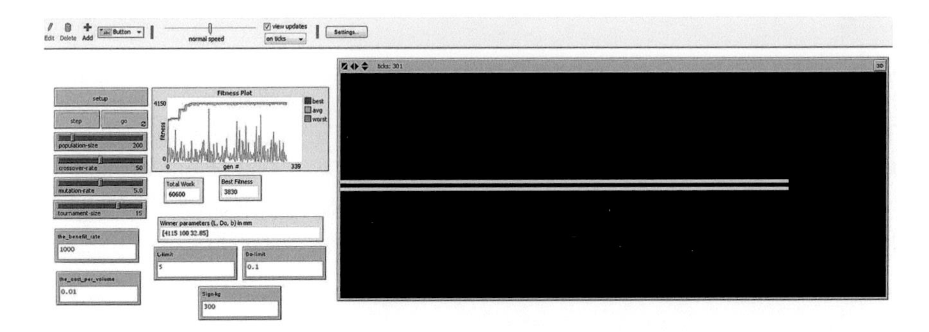

Fig. 13.6 GeneticAlgorithm_for_beam NetLogo model

Discuss It!

Run the GeneticAlgorithm_2D.nlogo model and see how the various controls affect the efficiency of reaching a solution (i.e., the amount of work), and how good the best-ever solution is compared to the ultimate best solution. Observe how the algorithm parameters (tournament size, mutation, cross-over, and population size) affect the ability of the algorithm to search the solution space and find the highest peak.

Answer It!

Q13.08: Using the GeneticAlgorithm_for_Beam.nlogo simulation, determine the *optimum* configuration assuming the benefit rate is $100 and the cost is $0.01 per volume. Limit tube length to a maximum of 5 m. Make sure the sign mass is 45 kg and a safety factor of 2. Does the placing of limits on the tube length (L-limit, tube length in meters) and tube thickness (Do-limit, tube diameter in meters) affect your optimum?

Q13.09: How much work (which is equal to the number of solutions computed) was done to determine your solution? Assuming that your three parameters (L, Do, and b) can vary in increments of 1 mm, how many possible solutions are there?

Discuss It!

You should have already thought about the importance of efficiently finding an optimum solution. Now think about the amount of work the two optimization algorithms used to solve the beam problem. Why do you think the Genetic Algorithm is an appropriate optimization algorithm for some types of problems and not others?

Discuss It!

You have only introduced two algorithms to solve optimization problems. There are many more that have been proposed and implemented based on other natural phenomenon. Think about some natural phenomenon in your discipline that might be used as an analogy for an optimization algorithm. Now, can you find a published paper that uses that analogy?

13.10 Linear Programming

A system of linear equations can be solved with linear programming which uses matrix algebra. Linear equations are those where the parameters are only in the first power. The objective function and all constraints must be linear, with each being either an equality (=) or an inequality (>, <, >=, <=). All parameters must be non-negative (sometimes you need to reformulate the equation to remove negativity).

An example of a simple two-parameter linear programming optimization problem is:

$$\text{maximize } O() = 30x + 20y \tag{4}$$

subject to:

$2x + y <= 500$
$x + y < 300$
$x < 400$
$x >= 0$
$y >= 0$

A Graphical Method can be used to solve this system, but is only applicable to two dimensions. A more general solution technique is the Simplex Method that was developed in 1947 by George Dantzig of Stanford University. It is an iterative algorithm using matrices that can find the optimum solution of an Objective Function for a large number of parameters and constraints. (See Chapter 16 of Burch 1994). The Simplex Method is introduced following.

13.11 Simplex Method

Following the description in Burch (1994), the process to solve an optimization problem using the Simplex Method proceeds with 10 steps:

1. *Define the decision variables.*
   In the example, that would be x and y.
2. *Define the objective function.* The function needs to be linear, include all the decision variables, and the direction (max or min) needs to be defined.
   (a) maximize $O() = 30x + 20y$
3. *Determine the constraints.* There are always limits on the parameters, and these constraints are represented mathematically.
   (a) $2x + y <= 500$
   (b) $x + y <= 300$
   (c) $x <= 400$
4. *Declare the sign restrictions.* Decision variables need to be positive, so if there isn't already a constraint that restricts the variable to the positive region, include it here.
   (a) $x >= 0$
   (b) $y >= 0$
5. *Convert the linear programming problem to standard matrix form* $[A]\{x\} = \{b\}$, where [A] is a matrix of all the equation variable coefficients, {x} is a vector of all the decision variables, and {b} is a vector of all the constants. Also, the objective function becomes $O() = \{c\}\{x\}$. Inequalities are removed by including slack variables, which are additional variables added to each equation. Note that our notation has changed from x and y to x_1 and x_2.
   (a) The equations for the constraints are:

$$2x_1 + x_2 + x_3 = 500$$
$$x_1 + x_2 + x_4 = 300$$
$$x_1 + x_5 = 400$$

6. *Prepare the first tableau*, which means to set up a table in a specific way so you can later manipulate to arrive at your solution. The tableau is of the form:

$$\begin{array}{ll} \{A\} & \{b\} \\ \{c^*\}\{A\} - \{c\} & \{c^*\}\{b\} \end{array}$$

where {c*} is a subset of the {c} vector with members who are the elementary columns (defined by one entry of 1, the rest zeros) in the tableau. The {c*}{A} – {c} entries are called the Simplex Indicators. The {c*}{b} is the current solution of the Objective Function.
   (a) For this example, the first tableau would be:

2	1	1	0	0	500
1	1	0	1	0	300
1	0	0	0	1	400
−30	−20	0	0	0	0

7. *Check to see if the current solution is maximal (optimal).* This is done by examining the Simplex Indicators – if none are negative, the solution is maximal and the algorithm stops. The solution can be read from the tableau (see Step 10).
8. *Check to see if no solution exists.* It is possible that your problem is not solvable (too constrained or unbounded). If there is a column that has all negative entries, no solution exists.
9. *Create a new tableau to find a better solution.* The new tableau is created by pivoting the matrix in the current tableau. Pivoting is accomplished as follows:
   (a) Pick the column with the most negative Simplex Indicator. Divide the last column with the values in the pivot column to find the lowest positive number – that will be the pivot row.
   (b) The pivot itself is just a row-reduction technique to end up with a 1 in the pivot row and zeros elsewhere in the pivot column.
   (c) The second tableau for our example is:

1	1/2	1/2	0	0	250
0	1/2	−1/2	1	0	50
0	−1/2	−1/2	0	1	150
0	−5	15	0	0	7500

10. Repeat Step 9 until either Step 7 or Step 8 is satisfied. When the algorithm is terminated with a solution, the optimized objective function value is given in the bottom right cell, with the right column indicating the optimal values for the variables (which variable is determined by the lone *1 s* in the columns).
   (a) In your example, the second tableau is does not meet Step 7 criteria, as there is still a negative Simplex Indicator. You pivot again (column 2, row 2) and get:

1	0	1	−1	0	200
0	1	−1	2	0	100
0	0	−1	1	1	200
0	0	10	10	0	8000

Since our Simplex Indicators are now all non-negative, you have arrived at a solution. The optimal value of our objective function is 8000, with x_1 (x) equal to 200 and x_2 (y) equal to 100. Recall that O() = 30x + 20y, so 30 * 200 + 20 * 100 == 8000. Checking to ensure that all the constraints have been satisfied:

2* 200 + 100 <= 500
200 + 100 <= 300
200 <= 400

The Simplex Method is a very efficient and powerful algorithm for solving Linear Programming problems.

Answer It!

Q13.10: It is highly likely that the Optimal Seating problem you defined above is a system of linear equations that can be solved using the Simplex Method. Do it. Are you sitting in your optimal location?

Q13.11: Is the Beam optimization problem described above solvable using the Simplex Method? What is the basis for your answer?

13.12 Computing Questions

- How reliable were the simulations you used/developed? Why?
- How valid where the simulations you used/developed? Why?
- How did you verify the simulations you used/developed?
- What were the possible sources are errors in your simulation(s)?
- Did your simulation(s) alleviate measure/time/cost and/or ethical issues? If so, how
- For the different integers in your module, what data size(s) do you think were used to represent them? Why?
- For the different floating point numbers in your module, what data size(s) do you think were used to represent them? Why?
- What are the different sources of errors that might be encountered in the computational work done in your module?
- Which of the data used in your lab was homogenous and which was heterogeneous?

- Explain one algorithm used by the computer while working on your module.
- For at least one algorithm in your module, what is the Big-Oh for that algorithm? Explain your answer. Would you be willing to double the size of the input?

13.13 Related Modules

- Module 3: Data Types: Representation, Abstraction, Limitations. Possible sources of errors.
- Module 5: Procedures: Algorithms and Abstraction. Introduction to algorithms.
- Module 9: Procedures: Performance and Complexity. Introduction to the issues related to complex problems and the potential inability for computing to solve even those problems.

References

Burch JG (1994) Cost and management accounting: a modern approach. West Publishing Company, Saint Paul. ISBN 0-314-02773-4

Stonedahl F, Wilensky U (2009) NetLogo simulated annealing model. http://ccl.northwestern.edu/netlogo/models/SimulatedAnnealing. Center for Connected Learning and Computer-Based Modeling, Northwestern University, Evanston, IL.

Data Organization and Analysis 14

14.1 Objectives

After completing this module, a student should be able to

- Describe an example of a spatial analysis problem.
- Define table, key field, record, attribute, query, join, cardinality, projection.
- Appreciate how GIS software can support spatial analysis.
- Construct correct attribute queries using SQL statements.

14.2 Definitions

- ArcGIS
- Attribute
- Cardinality
- Feature
- Feature Classes
- Geodatabase
- GIS
- Join
- Key (field)
- Precision
- Projection
- Query
- Raster model
- SQL
- Table
- Vector model

K. Brewer, C. Bareiss, *Concise Guide to Computing Foundations*,
DOI 10.1007/978-3-319-29954-9_14

14.3 Motivation

Information as a part of any science or engineering discipline requires appropriate organization to facilitate efficient and effective storage, use, and retrieval. This module will focus on spatial data, but the concepts presented could be applied to any organized data.

14.4 Spatial Data Representation

Spatial data can be represented in a computer information system by one of two models: raster or vector. The raster model is used for data that is best represented by a grid of discrete points (i.e. pixels). An example of raster data is an aerial photograph. The vector model is used for data that represents discrete objects (e.g. points, lines, areas) (Price 2011). Examples of vector objects could include fire hydrant locations, roads, and county boundaries.

Information about spatial objects is called attributes. Attributes can be added to any vector model data set and organized in an attribute table. The rows (records) of the table would represent each spatial object, and the columns (fields) are the attribute types. Note that location information for each spatial object is NOT an attribute, but just an essential part of the object's defining information.

Answer It!

Q14.01: For the data set of at least five northeast Illinois counties, create an attribute table with at least five attributes about each county. Use the Internet to find the data for each attribute for each county. (Don't forget that one important attribute of counties would be *name*!)

14.5 Joining Data

Often, one set of data you collect (or have) is related in some way to another data set – and you want to combine the data based on a common attribute. You were introduced to this type of *joining* in Module 10, but here we are going to expand on the concept of *joining*.

Below is an example of two tables of information. We will explore this familiar (and intuitive) example prior to looking at joining spatial data. The related information between the two tables is *Major*. Because they are the fields that allow a connection between the two tables, we call those the *key* fields (Tables 14.1 and 14.2).

After joining the University table to the Student table, the result is given in Table 14.3. (Note how the University data has been appended to the right of the

Table 14.1 Student data set attribute table

Object ID	First name	Last name	ID	Gender	Major
1	Jane	Doe	12321	F	Geology
2	Kelsey	Smith	23123	F	Chemistry
3	John	Will	83982	M	Biology
4	Kurt	Cook	30291	M	Engineering
5	Julie	Wang	76523	F	Math
6	Frank	Stanley	33321	M	Chemistry
7	Chris	Buck	54498	F	Biology

Table 14.2 University data set attribute table

Object ID	Major	Department	Required hours
1	Biology	Biology	128
2	Chemistry	Physical Sciences	132
3	Engineering	Engineering	137
4	Geology	Physical Sciences	128
5	Math	Math	129

Table 14.3 Join result table

Object ID	First name	Last name	ID	Gender	Major	Department	Required hours
1	Jane	Doe	12321	F	Geology	Physical Sciences	128
2	Kelsey	Smith	23123	F	Chemistry	Physical Sciences	132
3	John	Will	83982	M	Biology	Biology	128
4	Kurt	Cook	30291	M	Engineering	Engineering	137
5	Julie	Wang	76523	F	Math	Math	129
6	Frank	Stanley	33321	M	Chemistry	Physical Sciences	132
7	Chris	Buck	54498	F	Biology	Biology	128

Student data, and the University data is duplicated for each Student with the same Major.)

Answer It!

Q14.02: Identify which columns of which original tables (Table 14.1 or 14.2) are represented in the Table 14.3, and which columns are missing/omitted.

Going back to our Illinois county example, in addition to the county spatial data set you created, let's say you have a city data set with the following attribute Table 14.4:

Table 14.4 City data set attribute table

Object ID	Name	County	Pop1990	Elevation	Type
1	Kankakee	Kankakee	27575	663	City
2	Bourbonnais	Kankakee	13934	663	Village
3	Joliet	Will	76836	564	City
4	Chicago	Cook	2783726	596	City
5	Zion	Lake	19775	634	City

Table 14.5 Cardinality examples

			Source
		One	**Many**
Destination	**One**	One-to-One: County – County Seat (adding the County Seat information to County information. There is only one county seat per county.)	One-to-Many: County – Cities (adding city information to counties. There are many cities in each county. But each city is only in one county.)
	Many	Many-to-One: Cities – County (adding county information to city data. Each city is in only one county. So county information will go to many cities.)	Many-to-Many: Counties – Rivers (adding River information to Counties. There can be many rivers in each county, and rivers can go through multiple counties.)

Say you want to add the county information to the city information – in other words, you want to add the county attributes as extra attributes to each city record. You can join the city and county data sets easily because they have a common attribute: county name. The *name* attribute field in the county data set and the *County* attribute field in the city data set are the key fields and you can join based on them.

Answer It!

Q14.01: Before you join two data sets, you first need to think about the *direction* of the join: the *source* data set is the data you are going to add to the *destination* data set. What is the *source* data set in our above example? What is the *destination* data set?

The direction of the join isn't just an intellectual exercise, it can control whether the join is even possible. You can think of how a join is going to work by thinking of the *cardinality* – the relationship of the source and destination data sets. Cardinality terminology is ordered destination-to-source and can be one of four possibilities: one-to-one, one-to-many, many-to-one, and many-to-many. (See the following Table 14.5.)

One-to-one means that for each destination data set record, there is only one source data set record. One-to-many means that for each destination data set record, there are multiple source data set records, and so forth. Simple joins only will work for one-to-one or many-to-one.

Discuss It!
Can you think of examples for each cardinality type?

Answer It!
Q14.04: What will the cardinality be for your join to add the county information to the city information?

Discuss It!
If you join the city data set to the county data set, you would be adding the city attributes of Name, Pop1990, Elevation, and Type. What would the result be for these attributes for Kankakee county?

The example is small enough (five counties and five cities with five attributes each) that you can do this join by hand fairly quickly. But in real analyses you typically want to join data sets with many more records (and many more attributes). For example, the Census information for Illinois alone has 50 or more attributes for each of thousands of census tracts. To do this, you need to use GIS software.

Answer It!
Q14.05: Using GIS software (see Appendix for step-by-step instructions for one possible system) and the ILCounties and ILCities data sets (you can find these two data sets – and the two USA ones needed below – on the textbook webpage, along with download and installation instructions), perform the same join you did by hand. How long did it actually take to do the calculation/join? How many cities were used? How many counties? How many total attributes are in the joined attribute table?

Discuss It!
For the example above, you only used Illinois county and city data sets as it would be more complicated to do a join with USA counties and cities. Why?

Answer It!
Q14.06: You usually want to join data sets so you can get a deeper understanding of the data. Using your joined data, you could, for example, start to understand what cities are most important to their county. You could

assume that any city/county population ratio >50 % are most important. How could you easily calculate that from your joined data set? How many cities in Illinois have a city/county population ratio of >50 %?

14.6 Spatial Joining

Any database system could do the above analyses, as you didn't really use any spatial information for the objects (cities or counties). The power of Geographic Information Systems (GIS) is that it can use the spatial location to relate one object to others. This is where spatial analysis comes in – using location information to augment attribute analysis.

Answer It!

Q14.07: Again using GIS software and the USACounties and USACities data sets, join the city information to county information, but this time use the spatial location of the cities to determine what counties they are in (note there is not a county attribute in the cities data set). How long did it actually take to do the calculation/join? How many cities were used? How many counties? How many total attributes are in the final product?

Discuss It!

By using spatial locations in your joining, you don't need to have key fields; therefore, you don't need to know what key fields would be needed ahead of time! In your field of study, what spatial data might you collect that you might want to relate to another spatial data set?

14.7 Computing Questions

- For the different integers in your module, what data size(s) do you think were used to represent them? Why?
- For the different floating point numbers in your module, what data size(s) do you think were used to represent them? Why?
- What are the different sources of errors that might be encountered in the computational work done in your lab?
- Which of the data used in your lab was homogenous and which was heterogeneous?
- Explain one algorithm used by the computer while working on your module.

- What databases were used in your module? How many different tables were used for each example you studied? How many different fields in the tables did you use?
- For at least one algorithm in your module, what is the Big-Oh for that algorithm? Explain your answer. Would you be willing to double the size of the input?

14.8 Related Modules

- Module 3: Data Types: Representation, Abstraction, Limitations. Computational limits/errors (e.g. round-off, overflow, underflow, limited precision, non-reproducible computations) are examined.
- Module 5: Procedures: Algorithms and Abstraction. Algorithms and computation are examined further.
- Module 9: Procedures: Performance and Complexity. The impact of algorithm complexity with large datasets is introduced.

Reference

Price M (2011) Mastering ArcGIS, 5th edn. McGraw Hill, Dubuque. ISBN 978-0-07-336932-7

Appendix: NetLogo Tutorial

This appendix contains a quick introduction to programming in the NetLogo system. A more detailed set of instructions is available at:

https://ccl.northwestern.edu/netlogo/docs/

General Things

Let us begin learning about NetLogo programming by trying some built-in commands. First, select the *File* menu and then select *New*. Your text cursor should be in a panel near the bottom called *Command Center*. The text cursor should be flashing in a text area with the word *observer* next to it. NetLogo programming commands are given to various types of agents that can respond to the commands. The main two types of agents we are concerned with right now are the *observer* and *turtles*. To change the agent type you want to give commands to, you can select the type by selecting it with the mouse. You can also use your *tab* key to move through the agent types. Try both methods now.

Now click on the *settings* button near the top right of the screen. Notice that the model world is a 2-dimensional grid of non-movable agents called *patches*. The origin (0, 0) is in the center by default with the x-axis going from −16 to 16 and the y-axis is the same. Turtles are movable agents that can move around the world. Click *OK* to save the world settings.

Built-In Commands, Turtles, and Variables

Try your first built-in command. Type the observer command ***show 2+2*** in the command center and press *enter*. You should see a response from the observer agent in the command center text area showing the calculated answer *4*. NetLogo already knows about numbers and how to add them. Try other basic arithmetic. Hint: the basic arithmetic operators are +, −, *,/.

Now try the command ***create-turtles 1***. This asks the observer agent to create one movable turtle with a random color and heading, default shape, and location. Do you see a pointer near the center of the world? Change the agent type to *turtles*.

K. Brewer, C. Bareiss, *Concise Guide to Computing Foundations*,
DOI 10.1007/978-3-319-29954-9

Give your turtle the command ***set color red***. Do you see a change in your turtle's color? Try changing the turtle to another color.

If you want to clear the world and start over, you can give the observer agent the command ***clear-all***. Try it now. Did your turtle disappear? Before continuing, create 1 turtle in the world.

Change the agent type to *turtles*. You should only have one turtle in the world right now, but these commands are sent to all turtles in the world. You'll try more turtles later. Each turtle has a set of variables with values specifically for that turtle. You can see the variables and their values for your turtle with the command ***inspect turtle 0***. You can also get this information by clicking on *Tools* menu and selecting *Turtle Monitor*. Another way to monitor a turtle is to right click the mouse on a turtle and select *inspect turtle* from the pop-up dialog.

Values of variables associated with agents can be changed using the *set* command. You already changed the *color* variable of your turtle for example with the command *set color red*. You just give the agent the command (*set*) followed by the variable you want to change (*color*) and the value you want to change it to (*red*). Since turtles are movable agents, they can change their world grid coordinates (xcor, ycor). You can move the turtle to grid location (2, 2) with these two commands ***set xcor 2 set ycor 2***. There is also a shorter alternate command ***setxy 2 2*** that does the same thing. Try moving the turtle to location (2, 2). Did it work? Try moving the turtle to other locations.

To save time typing in the command center, you can use the up and down arrow keys to go back over command history and edit commands before pressing enter to run the command. Use the arrow keys to go back to a previous command and change it slightly before running it again. Did you get the results you expected in the world?

The direZction a turtle is "heading" is represented by a *compass heading* as a number of degrees from 0 to 359. Remember that 0 = north, 90 = east, 180 = south, 270 = west. Note: this is not the same as the normal orientation used in geometry and trigonometry which starts at 0 to the right and increases going counter-clockwise. Change your turtle's heading to east. Your turtle should now be heading toward the right of the world, which is east. Try changing the turtle heading to other directions. Think about what the result should look like before you run the command. Then run it and see if you get the result you expected.

You've already changed the turtle's *absolute* world grid location and heading. You can also change the turtle's *relative* location and heading. Each turtle knows its grid location and heading. Giving a turtle the command ***forward 4*** tells the turtle to move straight ahead 4 patches. You can also change the heading from the current heading with ***right 90***. This makes a 90 degree or right turn relative to the current heading. Try making some turns and moving around in the world to get a feel for controlling the turtle.

Each turtle has a pen that can be raised or lowered to leave a mark where the turtle has been. These turtle commands are ***pen-up*** and ***pen-down*** respectively. Try putting the *pen-down* and making a square that is 10 patches by 10 patches in the world grid. Carefully plan the steps the turtle must make to move along each side of the square. Then try to give the sequence of commands to make the turtle complete the square.

Repetition

Often you can shorten a procedure that *repeats* a sequence of commands. The simplest repetition control structure in NetLogo is ***repeat*** which repeats a fixed number of times on a sequence of commands. Change the color of your turtle and then try this command ***repeat 4 [forward 10 right 90]***. *Forward 10 right 90* is a *sequence* of commands which must be enclosed in brackets. This sequence of 2 commands is repeated 4 times. The computer must still run all 8 commands, but it is shorter for you to give the command. It achieves the same result as writing out all 8 commands as a single sequence.

Notice that after the turtle draws a 10 x 10 square, it is heading in the same direction as it was before drawing the square. Give the ***right 5*** command. Now use the command editor to run the command that draws a square again. Repeat these two commands several times and watch the pattern that is developing.

With the command editor, repeat a sequence of commands the number of times you just calculated. The sequence of commands in brackets should include the commands to draw a square followed by a command to turn right 5 degrees. Does it look like you planned? If not, try to figure out why.

Selection

NetLogo has the command ***ifelse*** for selecting between 2 sequences of instructions depending on some condition that is true or false. To see how it works, clear everything and create one turtle. Now try this command for the turtle: ***ifelse heading mod 10 < 5 [set color yellow] [set color green]***. This command checks the condition *heading mod 10 < 5*. This divides the turtle's current heading by 10 and checks to see if the remainder is less than 5. Since the remainder of dividing any number by 10 must be between 0 and 9, headings that end between 0 and 4 degrees will be true and those that end between 5 and 9 will be false. When the condition is true, the turtle color will be set to yellow. When it is false, the color will be set to green. Did it work correctly? Change the heading using the ***set heading*** command. Then use the command editor to run the same ifelse command and see if it does something different. It only does one of the ***set color*** commands when it runs.

Procedures

A reusable set of instructions, a procedure, can be created in the Code tab. The basic structure of a procedure is:

```
to procedureName
  do the necessary procedure commands here
end
```

Say we want to have a turtle draw a square. The movement of a turtle in a square would be ***repeat 4 [forward 10 right 90]*** and the turtle is right back where it started. Say we associate the name ***draw-square*** with this procedure:

```
to draw-square
  pen-down
  repeat 4 [forward 10 right 90]
  pen-up
end
```

To use a procedure, just write the name. For the above turtle procedure to draw a square, the code could be:

```
Clear-all
Create-turtles 1
ask turtles [ draw-square ]
```

Setup and Go

To common procedures in all NetLogo models are *setup* and *go*. *Setup* is a procedure that is expected to be executed once to "setup" the model. The *go* procedure is expected to be executed over and over until the user stops the model. These procedures are connected to buttons on the interface. The two buttons should be implemented as follows (Fig. A1):

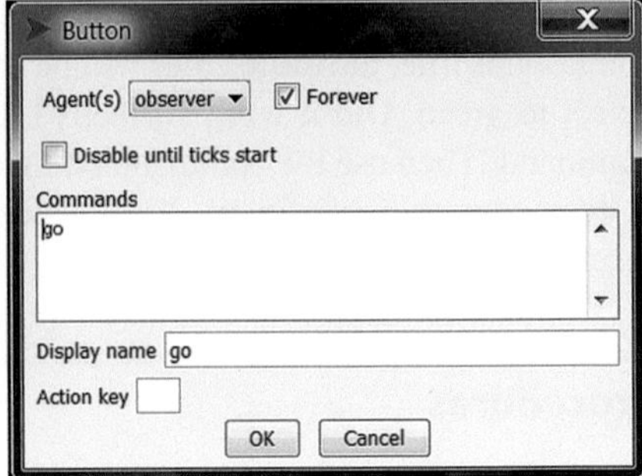

Fig. A1 Setup and go button implementation

A Basic Model

A basic, simple model is therefore:

```
;; A simple 1st NetLogo program

to setup
 clear-all
 create-turtles 1
 reset-ticks
end

to go
 ask turtles [
 draw-square
 set heading heading + 10
 ]
end

to draw-square
  pen-down
  repeat 4 [forward 10 right 90]
  pen-up
  set color random 20
end
```

Appendix: LabQuest Tutorial

This appendix contains step-by-step instructions on how to use the LabQuest units. A more detailed set of instructions is available from the manufacturer at:

http://www.vernier.com/files/manuals/labquest_quickstart_guide.pdf

Overview

The LabQuest instrumentation system consists of a LabQuest unit, sensors, and the LabQuest software. The system has been designed for educational/classroom use, so it has been made flexible and simple. The LabQuest units have a pressure sensitive touch-screen interface that works best using the attached stylus.

Additional help and explanation of the system can be found at the manufacturer's website: http://www.vernier.com

Adding Sensors

Sensors are *plug and play* and either plug-in on the side or top of the LabQuest unit (depending on the sensor plug—the two kinds are similar, so if a sensor plug doesn't fit in the top, try the side...). Since the sensors are *plug and play*, no setup or device management configurations are needed for the LabQuest to recognize and start collecting data from a sensor. Sensors can be added or removed at any time. Up to 6 different sensors can be attached at one time.

LabQuest Interface Overview

The software interface is a set of screens that provide control and display of attached sensors. There are three main screens that can be accessed by pressing the respective icons at the top of the display.

K. Brewer, C. Bareiss, *Concise Guide to Computing Foundations*,
DOI 10.1007/978-3-319-29954-9

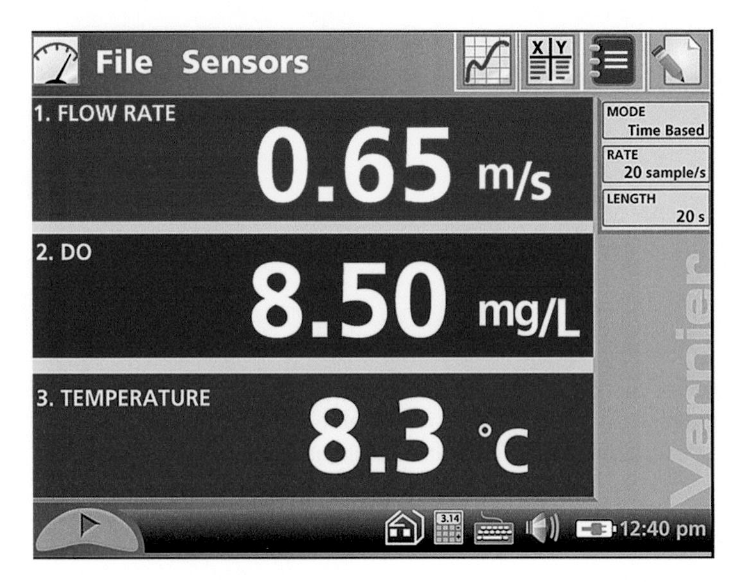

Meter view

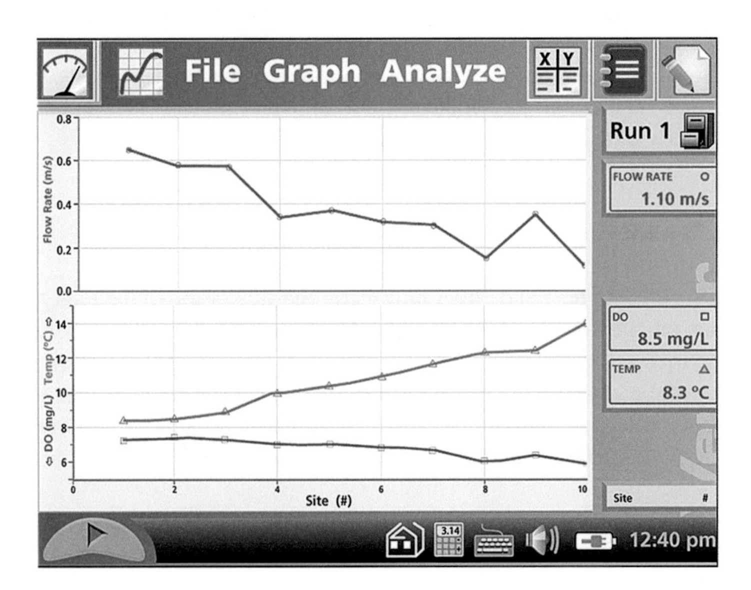

Graph view

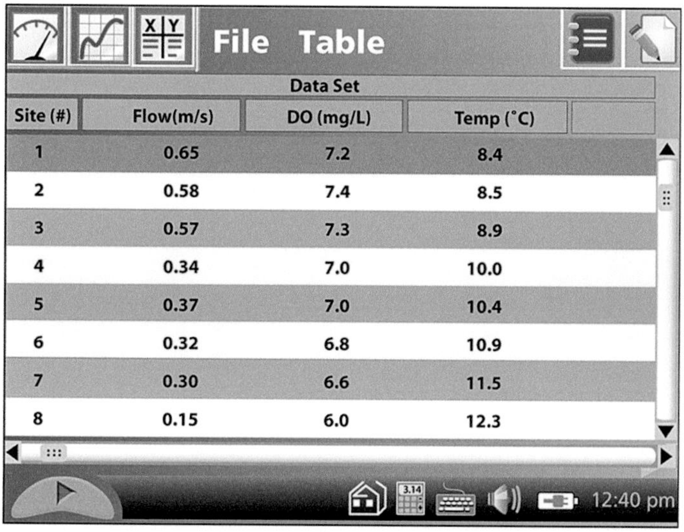

File Table

Data Set

Site (#)	Flow(m/s)	DO (mg/L)	Temp (°C)
1	0.65	7.2	8.4
2	0.58	7.4	8.5
3	0.57	7.3	8.9
4	0.34	7.0	10.0
5	0.37	7.0	10.4
6	0.32	6.8	10.9
7	0.30	6.6	11.5
8	0.15	6.0	12.3

3.14 12:40 pm

Datatable view

The Meter View is where you can change the data collection settings for the sensors, and can reset zero values and calibrate the sensors (for some sensors). The Graph View is where you see your collected data, and where you can interact with a graph of your data and perform simple analyses. This is also the view where you will be able to export your data. The Datatable View shows your data in tabular format. You can also export your data from this screen, if desired.

Changing Sensor and Data Collection Settings

From the Meter View screen, the *Sensors* menu will show all available actions for your attached sensors.

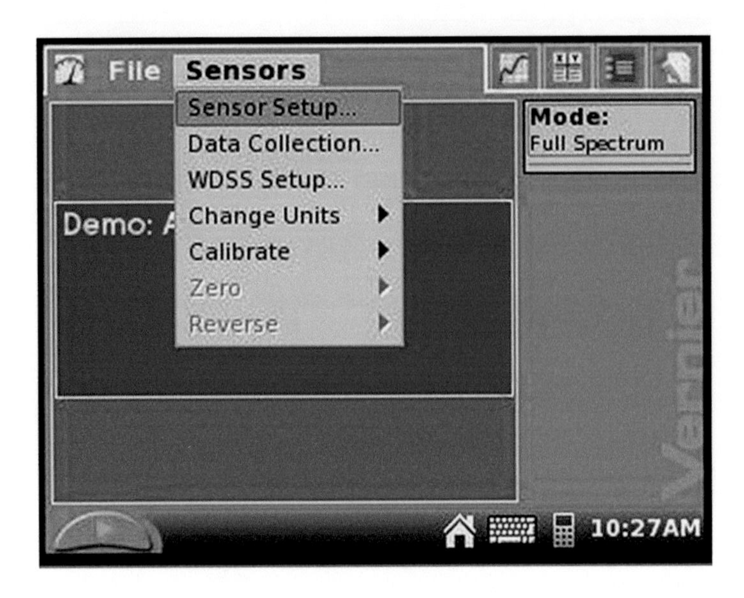

1. *Sensor Setup...*—you will rarely need to setup a sensor as this is automatically handled when you plug the sensor in. Use this only if you think your sensors are not behaving properly (or just unplug and plug them back in).
2. *Data Collection...*—this is where you can change how data will be collected via the sensor. Out of the many options for each sensor, you will likely just change the frequency and length of time for your data collection. Note that there is a maximum collection rate and maximum data values that can be collected, which may limit your frequency/sample time choices.
3. *Change Units*—this will allow you to collect data in different (but appropriate) units.
4. *Calibrate*—this will allow you to calibrate your sensor, if necessary. This is often not needed.
5. *Zero*—this will allow you to set the *zero* value for your sensor.
6. *Reverse*—this will allow you to reverse the numeric response of the sensor.

Collecting Data

After sensors are plugged in, they immediately begin "sensing" and displaying values on the Meter View screen. To collect a dataset, you need to press the *collect* button (that looks like it has a *play* symbol, go figure). After a brief initialization period, the LabQuest will begin collecting and displaying data in the Graph View. If you want to stop data collection before the end of the Sample Time, you can press the *collect* button again. Data collection automatically stops at the end of the Sample Time.

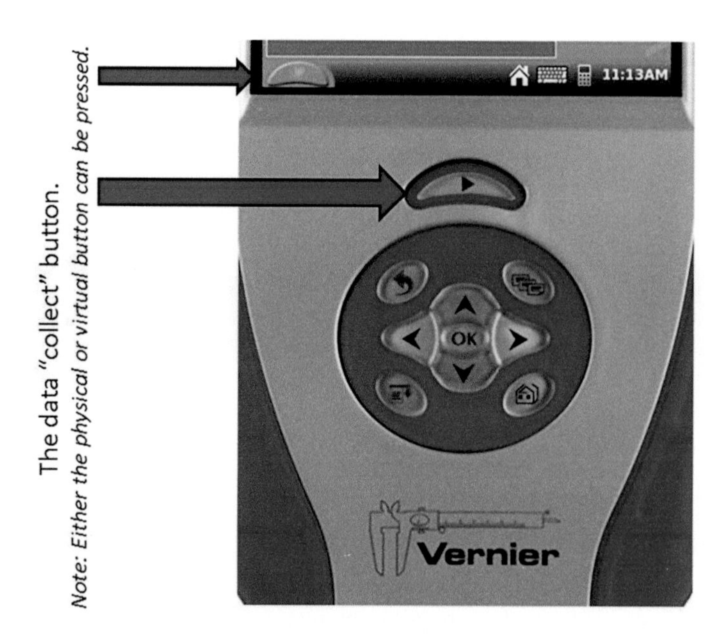

The data "collect" button.
Note: Either the physical or virtual button can be pressed.

Analyzing Data

After you collect data, you can use the LabQuest unit to perform some simple analyses. One of the most common techniques is to obtain simple statistics about your data. From the Graph View screen, select *Statistics* from the *Analyze* menu and then select (via check box) the data you wish to analyze (in the following example, it is Sound Pressure).

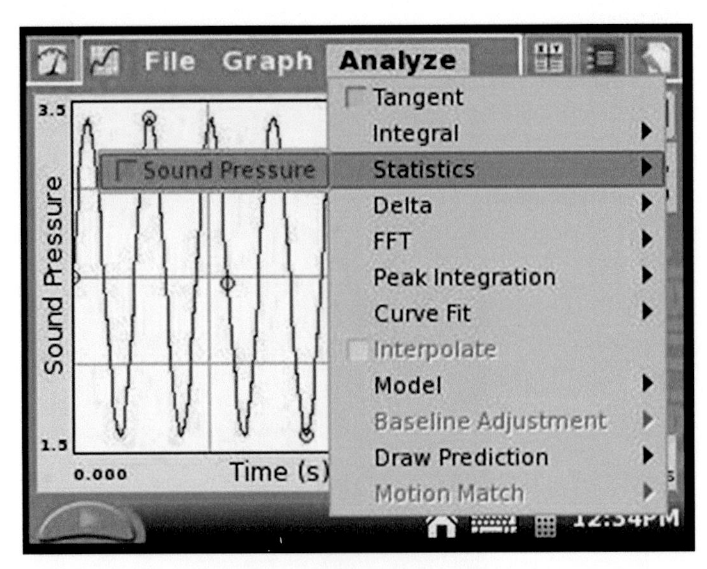

To the right side of the graph, the light blue box will display basic statistics on your data. If you click that box, it will bring up a full screen view.

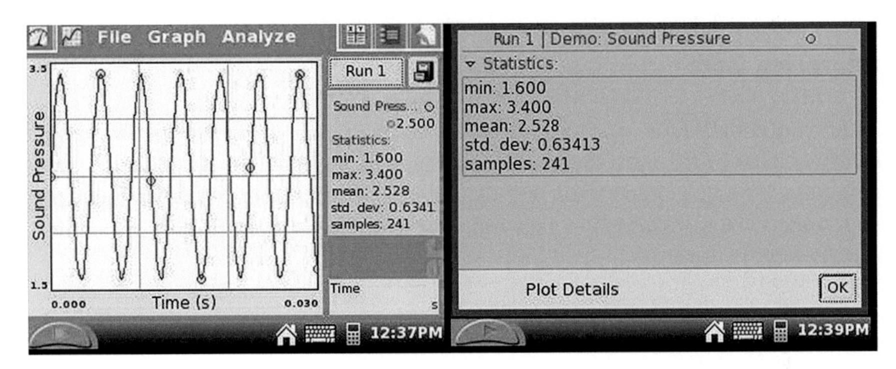

Getting Data to Excel

Once you have collected your data, you can export it to your computer to use in Excel (or other program). You will need a USB thumb drive. Follow the following steps:

1. Plug in your USB thumb drive in the USB port at the top of the LabQuest.
2. From the Graph View or Datatable View screen, select *Export...* from the *File* menu.

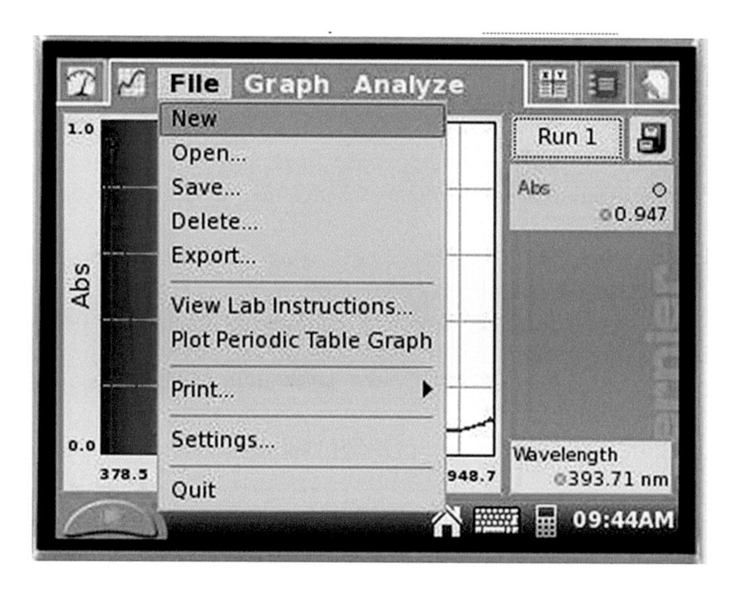

3. Tap on the USB thumb drive icon.

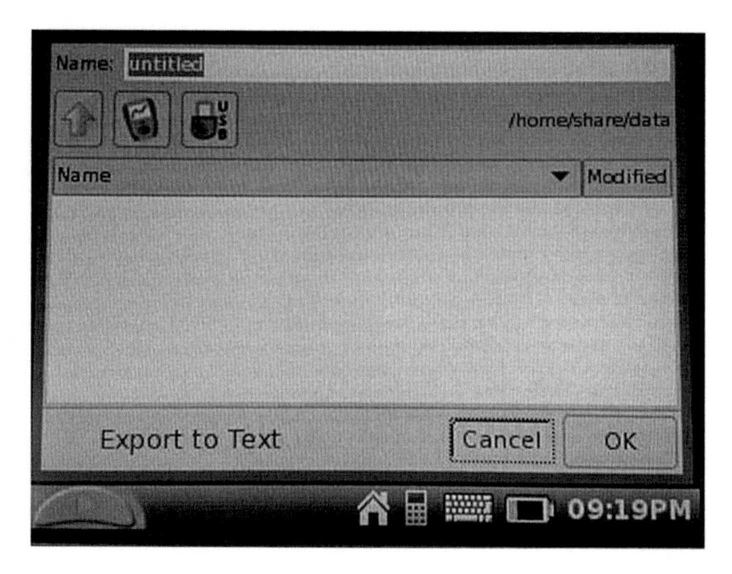

4. Tap the Name: field and enter the filename for your data. Use.*txt* as your file extension.
5. Tap *OK* and your data will be saved.
6. When finished (your USB thumb drive light stops flashing), un-plug your thumb drive.

7. Plug your thumb drive into your USB port on your computer.
8. Open your file in your computer's file system viewer (Windows Explorer on Windows; Finder on MacOSX). Move your file to the working directory of your choosing.
9. Open Excel.
10. Select *Open* from the *File* tab/menu. Change the file type to *All files (*.*)*.
11. Open your data file.
12. In the *Text Import Wizard—Step 1 of 3,* choose *Delimited Row 1* and *UTF-8* (these may be the default selection). Click *Next.*
13. In the *Text Import Wizard—Step 2 of 3,* choose *Tab* (this may be the default selection). Click *Next.*
14. In the *Text Import Wizard—Step 3 of 3,* leave all the default settings and click *Finish.*
15. You are now viewing your data in Excel, column format. Each data point is a separate row.

Reference

Software from Vernier Software & Technology and LabQuest App

Appendix: GIS Tutorial

This appendix contains step-by-step instructions on how to use the gvSIG version 1.11 GIS software. More recent versions of the software may behave differently.

Join by Attribute

1. Start gvSIG.
2. Click on the *View* icon.
3. Click on the *New* button. You should now see:

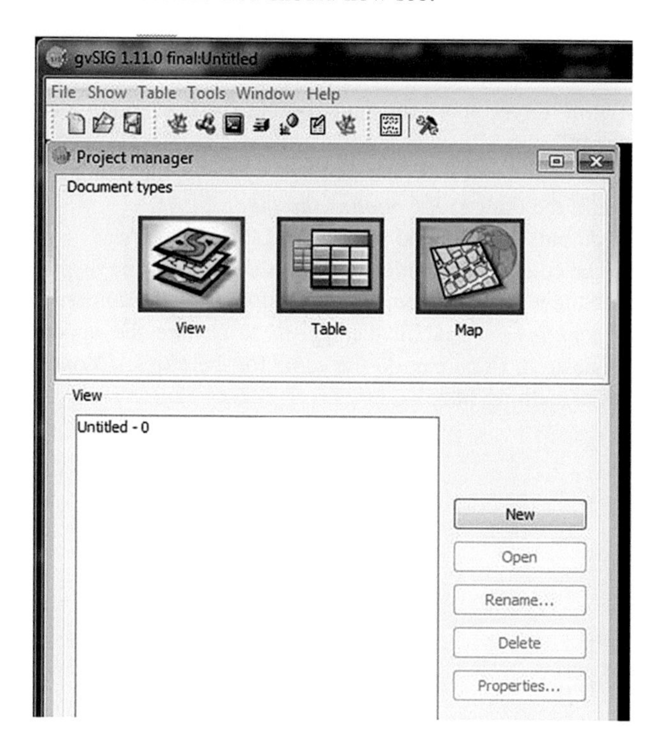

K. Brewer, C. Bareiss, *Concise Guide to Computing Foundations*,
DOI 10.1007/978-3-319-29954-9

4. Double-click the *Untitled –0* item in the list. This will bring up the following window:

5. Click the *Add Layer* icon (or find the *Add Layer* menu item under the *View* menu).
6. Click the *Add* button and from the standard windows file dialog, select (use the *open* button in the dialog) *ILCounties.shp file*.
7. Click the *Add* button again and select the *ILCities.shp file*.
8. Click *OK*. The counties and cities in Illinois should now be displayed in a map view. Resize the window to your liking. Double click the colored rectangle just below the *counties.shp* text in the *legend* to change the appearance of the counties, if desired. (You can do the same for the cities.). Your map/window should look something like:

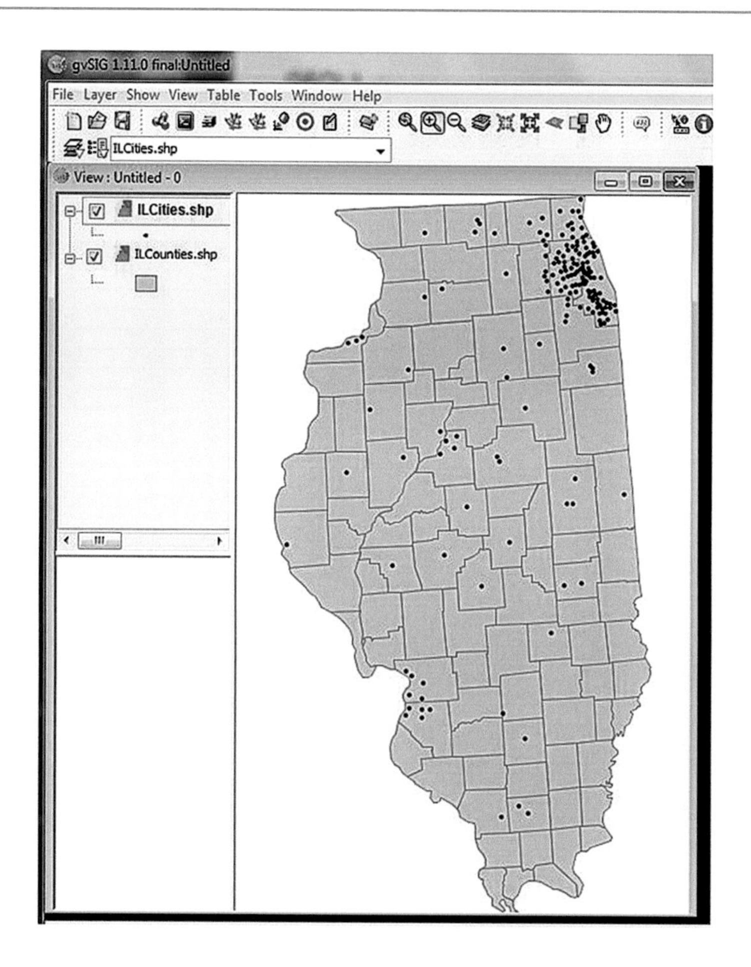

9. Explore the attributes for both the cities and counties. Click on *ILCities.shp* in the legend (the typeface will become **bold**, indicating it is the active layer) and then click on the *Show attribute table of active layer* icon in the toolbar. Resize the resulting window as required.

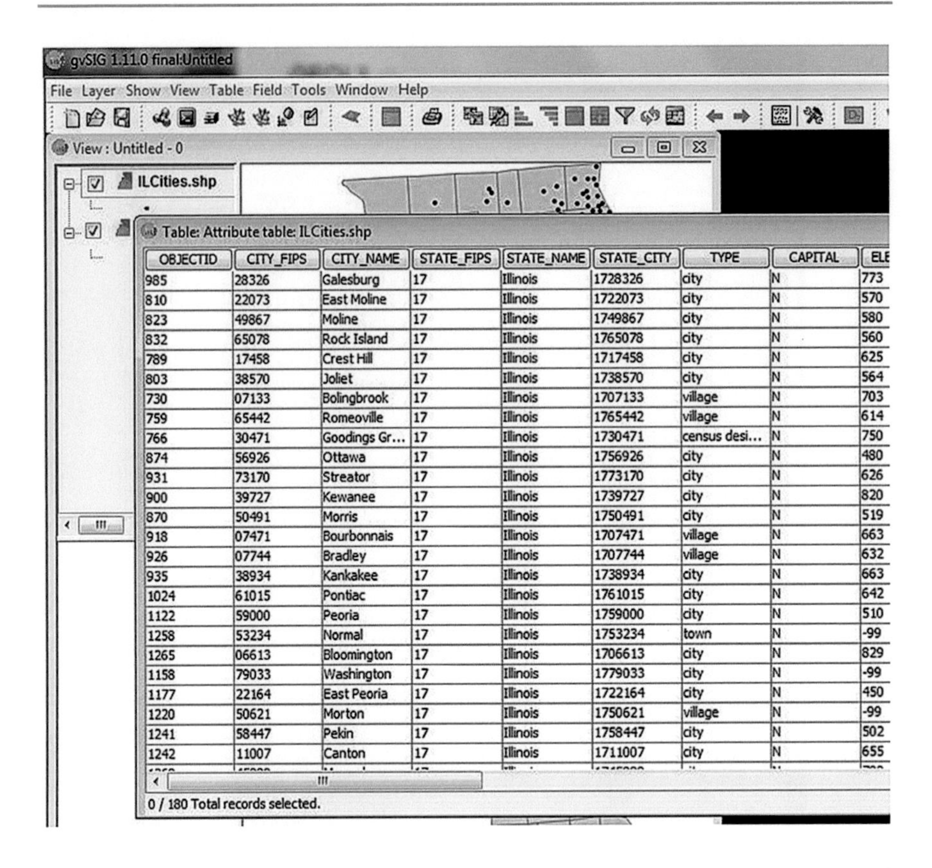

OBJECTID	CITY_FIPS	CITY_NAME	STATE_FIPS	STATE_NAME	STATE_CITY	TYPE	CAPITAL	ELE
985	28326	Galesburg	17	Illinois	1728326	city	N	773
810	22073	East Moline	17	Illinois	1722073	city	N	570
823	49867	Moline	17	Illinois	1749867	city	N	580
832	65078	Rock Island	17	Illinois	1765078	city	N	560
789	17458	Crest Hill	17	Illinois	1717458	city	N	625
803	38570	Joliet	17	Illinois	1738570	city	N	564
730	07133	Bolingbrook	17	Illinois	1707133	village	N	703
759	65442	Romeoville	17	Illinois	1765442	village	N	614
766	30471	Goodings Gr...	17	Illinois	1730471	census desi...	N	750
874	56926	Ottawa	17	Illinois	1756926	city	N	480
931	73170	Streator	17	Illinois	1773170	city	N	626
900	39727	Kewanee	17	Illinois	1739727	city	N	820
870	50491	Morris	17	Illinois	1750491	city	N	519
918	07471	Bourbonnais	17	Illinois	1707471	village	N	663
926	07744	Bradley	17	Illinois	1707744	village	N	632
935	38934	Kankakee	17	Illinois	1738934	city	N	663
1024	61015	Pontiac	17	Illinois	1761015	city	N	642
1122	59000	Peoria	17	Illinois	1759000	city	N	510
1258	53234	Normal	17	Illinois	1753234	town	N	-99
1265	06613	Bloomington	17	Illinois	1706613	city	N	829
1158	79033	Washington	17	Illinois	1779033	city	N	-99
1177	22164	East Peoria	17	Illinois	1722164	city	N	450
1220	50621	Morton	17	Illinois	1750621	village	N	-99
1241	58447	Pekin	17	Illinois	1758447	city	N	502
1242	11007	Canton	17	Illinois	1711007	city	N	655

10. At the bottom of the table view, the total number of records are given for that layer. The above image shows that there are 180 cities in the ILcities.shp layer. (How many counties are there in the ILCounties.shp layer?)
11. Before you join the county data to the city data using attributes, you need to know what the key fields are named in each layer. After you determine that, you can proceed.
12. To join based on attributes, with the attribute tables for both the ILCities.shp layer and ILCounties.shp layer visible, click the *Join...* icon.
13. The first entries are for the destination table.

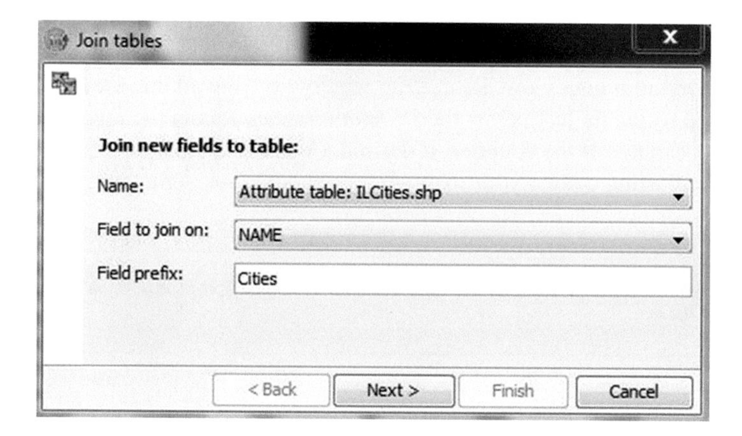

14. After clicking *Next*, the second entries are for the source table.

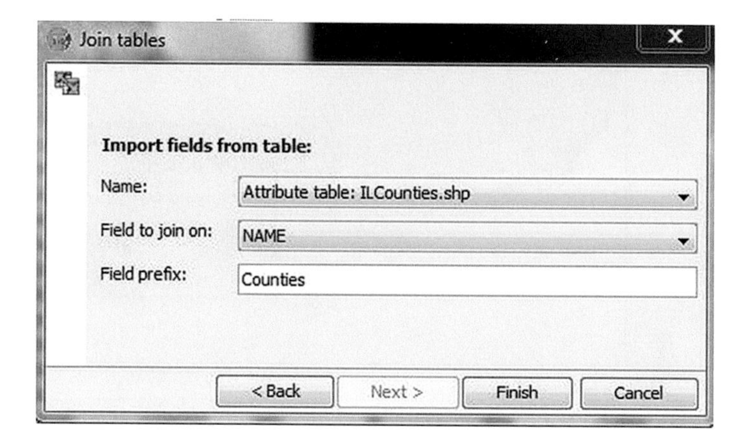

15. Click *Finish* and your ILCities.shp attribute table now has the appropriate county data fields added for each city record.

Spatial Join

1. Start gvSIG.
2. Click on the *View* icon.
3. Click on the *New* button.
4. Double-click the *Untitled –0* item in the list.
5. Click the *Add Layer* icon (or find the *Add Layer* menu item under the *View* menu).
6. Click the *Add* button and from the standard windows file dialog, select (use the *open* button in the dialog) **USACounties.shp** file.

7. Click the *Add* button again and select the **USACities.shp** file.
8. Click *OK*. the counties and cities in the lower-48 states should now be displayed in a map view. Resize the window to your liking. Double click the colored rectangle just below the *USAcounties.shp* text in the *legend* to change the appearance of the counties, if desired. (You can do the same for the cities.). After adjusting colors, your map/window should look something like:

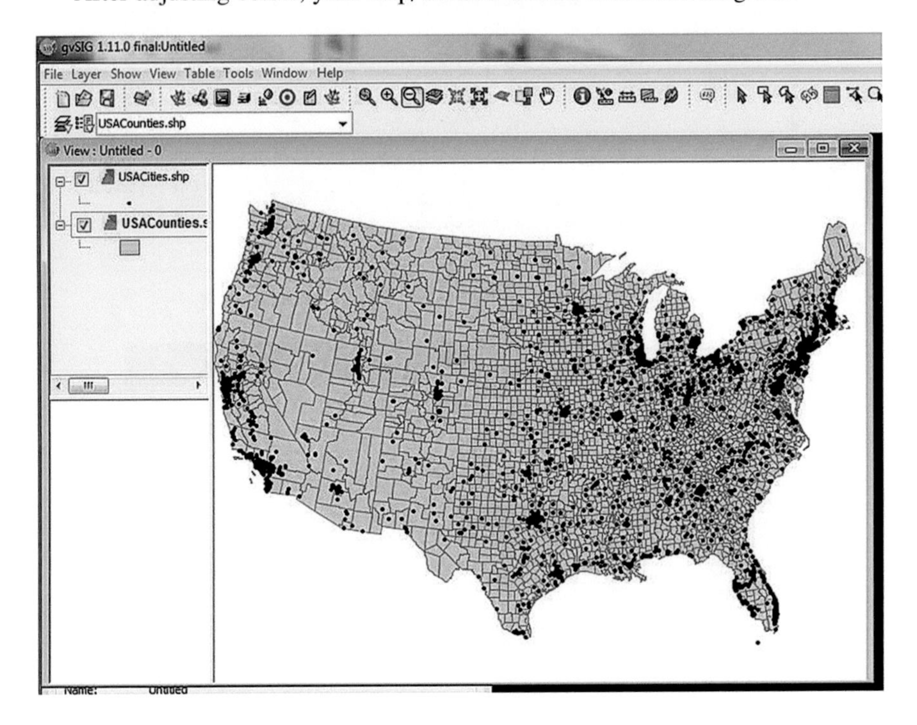

9. Click on the *Geoprocessing Tools* icon and expand the resulting tree to show the *Spatial join* tool:

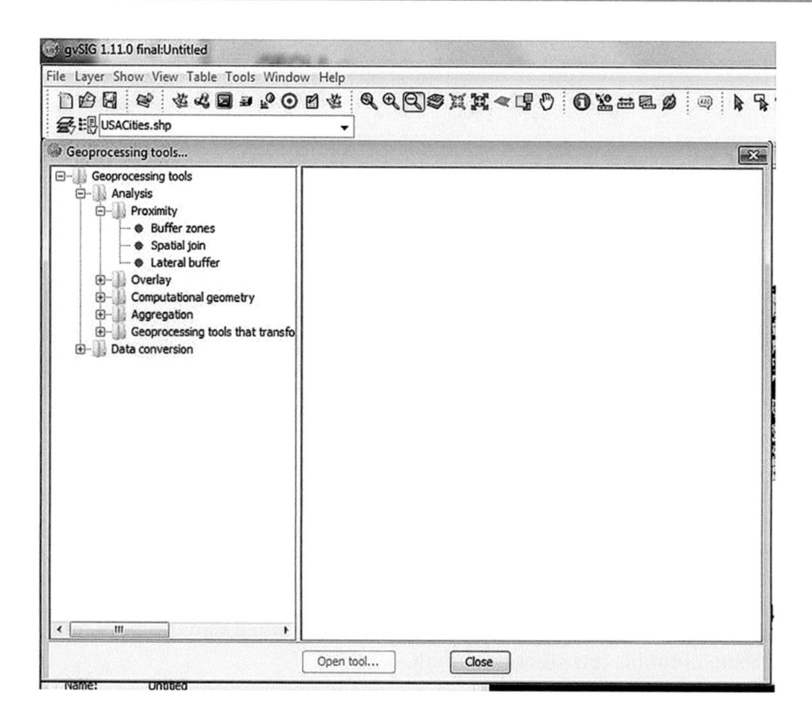

10. Click the *Spatial join* tool to bring up its description. Click the *Open tool...* button to open the tool's dialog box. The Input layer is the destination layer, and the overlay layer is the source layer. Choose a file name and location to save your joined layer.

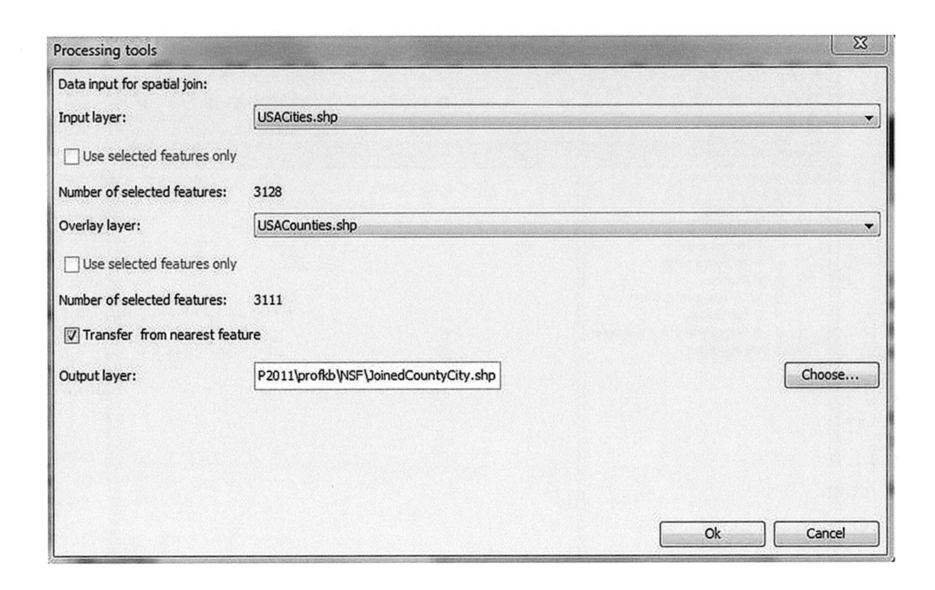

11. Open the attribute table for the joined layer and notice that the right half of the attribute columns are all county data.

Definitions

(A)

- *Abstraction*—general concept created from specific examples.
- *Active site*—pocket or cleft in a protein which is surrounded by amino acid side chains that help to bind the substrate and other side chains that play a role in the catalytic process.
- *Accuracy*—closeness of instrument-reported values to actual measurement values.
- *Agent-based modeling*—representation of a system–as a collection of autonomous decision-making agents that individually assess their situations and make decisions based on a set of rules.
- *Algorithm*—set of instructions that solves a particular problem.
- *Amino acid*—group of organic compounds containing both carboxyl and amino groups. There are 20 different kinds of amino acid molecules, each of which are commonly represented by a unique capital letter.
- *Anisotropic*—a property or parameter that varies depending on the direction of measurement
- *ArcGIS*—industry-standard GIS program sold by ESRI.
- *Array*—data structure with homogeneous data; each item can be directly addressed by a unique id that is sequential.
- *ASCII*—American Standard Code for Information Interchange. A standard for mapping single characters to an unique 7 bit number
- *Attribute*—information about a spatial object; multiple attributes are stored in fields in attribute tables.

(B)

- *Big-Oh classes*—common simplified functions that are upper bounds for actual *time complexity functions* in which algorithms are classified as "On the Order of" a group.
- *Bimolecular*—type of reaction initiated when two molecules collide.
- *Binary*—Base 2 number system whose only allowed digits are 0 and 1
- *Binary tree*—tree in which each node may have, at most, two children (a left and a right child).
- *Bit*—a single digit in binary

K. Brewer, C. Bareiss, *Concise Guide to Computing Foundations*,
DOI 10.1007/978-3-319-29954-9

- *Boundary conditions*—the definition of how the unknown value is defined at the boundaries of the domain. There are three main types of boundaries (Dirichlet, Neumann, and Cauchy).
- *Byte*—8 bits

(C)
- *Calibration*—the alteration of numerical model parameters to minimize error of numerical model results to actual (known) data.
- *Cardinality*—relationship between two data sets based on some commonality. In ArcGIS, x -to- y cardinality refers to the *destination* -to- *source* data. (Possibilities are: *one* -to- *one*, *one* -to- *many*, *many* -to- *one*, and *many* -to- *many*.)
- *Cauchy Boundary Conditions*—flux across the boundary is dependent on the unknown variable value.
- *Chain*—one portion of a protein
- *Compression*—process of taking a file and making it smaller so that the original (or close approximation of the original) can be restored.
- *Computational Science*—field of study concerned with using computers to run numerical calculations of mathematical models to analyze and solve scientific problems.
- *Concentration*—amount of solution per unit volume. In Chemistry, concentration is often measured as moles of solute per liter of solution.
- *Constraints*—equations or limits on parameters due to physical or other reasons.
- *Convergence*—the process of reducing the error to a minimum during an iterative process.
- *Curve fitting*—the process of finding the best curve to fit a set of data points

(D)
- *Database*—collection of self-describing integrated records.
- *Data set*—group of related data.
- *Data structure*—system to organize data in a particular way.
- *DBMS*—Database Management System, software that maintains a database.
- *Decimal*—Base 10 number system with digits from 0 to 9
- *Design*—the process of deciding how an algorithm or simulation should work before coding it
- *Dirichlet Boundary Conditions*—the unknown variable is known and defined at the boundary.
- *Discritization*—is process of creating a finite element or finite difference grid in a domain.
- *Domain*—is the area (or volume) of interest that is constrained by boundary conditions.
- *Dominant term*—fastest growing term in a *time complexity function* and is used to determine the *Big-Oh* class.

- *Drift*—tendency for reported instrument values to slowly change over time (when the actual measurement isn't changing), commonly due to electronics issues in the instrument and sensor.
- *Dynamic systems modeling*—representation of a system using feedback loops and stacks and flows.

(E)

- *Equilibrium*—state in which a reaction appears not to progress any further. Concentrations of the reactants and products reach a dynamic state in which the forward and reverse reactions are equal.
- *Expectation value*—average value of an observable property of a system measured once on many identically prepared experimental models.

(F)

- *Feature*—spatial object that uses either the raster or vector model.
- *Feature Classes*—grouping of like features in a GIS system. (Also called *Data Sets.*)
- *Finite difference*—a numerical technique to solve partial differential equations in a domain using algebraic equations and a grid.
- *Finite element*—a numerical technique to solve partial differential equations in a domain using a series of interpolation functions and a triangular grid.
- *Floating point*—integer followed by a decimal point and an integer.

(G)

- *Genetic algorithm*—optimization technique/method based on an analogy of genetics and evolutionary theory/natural selection.
- *Geodatabase*—object-oriented model for storing spatial information used by ArcGIS. Constructed on the architecture of standard relational database systems (for small, personal geodatabases, ArcGIS uses Microsoft Access). Contains feature classes (i.e., spatial objects of a similar type), tables, relationships, network information. A "robust" implementation, topology (i.e., relationship rules between spatial objects), network (i.e., connectedness), and behavior and validation rules.
- GIS—Geographic Information System,
- *Graph*—data structure with nodes that are connected with edges.

(H)

- *Heme*—prosthetic group that consists of an iron atom in the center of a large, heterocyclic organic ring called a porphyrin.
- *Heterogeneous*—a property or parameter that varies in space
- *Heuristic search*—A search method to find a satisfactory solution using shortcuts or preexisting experience. Examples include using a rule-of-thumb or educated guess to shorten the search time or minimize required resources (e.g. storage). The search is not guaranteed to be complete.
- *Hexidecimal*—Base 16 number system with digits from 0–9 and a-f
- *Homogeneous*—a property or parameter that does not vary in space

(I)

- *Integer*—any whole number including 0 and negative numbers.
- *Isotropic*—a property or parameter that does not vary depending on the direction of measurement
- *Iteration*—the repetition of a solution algorithm/method to arrive at a converged solution.

(J)

- *Join*—combination of two tables based on common attributes or spatial relationship.

(K)

- *Key (field)*—common field that is used when combining tables.
- *Kinetics*—study of motion with respect to time. In Chemistry, the term refers to the study of the progress of reactions over time and the mechanisms through which reactions proceed.

(L)

- *Ligand*—ion or molecule attached to a metal atom by bonding in which both electrons are supplied by one atom.
- *Linked list*—data structure with homogeneous data in which each item knows where the next item is.
- *Lossless compression*—compression method that guarantees the exact restoration of the original.
- *Lossy compression*—compression method that does not guarantee the exact restoration of the original.

(M)

- *Markup language*—language that uses tags to describe its contents.
- *Model*—mathematical representation of some scientific phenomenon or system.

(N)

- *Neumann Boundary Condition*—flux across the boundary is known and defined.

(O)

- *Objective function*—mathematical function designed to represent a system.
- *Optimization*—process of finding the best solution.

(P)

- *Parameter*—independent variable
- *Particle in a box*—type of mathematical problem used as a simple model for quantum mechanics problems.

- *Performance*—inverse proportion of the time required to run a procedure on a computer system. shorter run time means higher performance.
- *Pixel*—single addressable dot on a graphical computer display.
- *Polypeptides*—short chains (50 or fewer) of amino acids linked by peptide bonds.
- *Potential energy well*—region in which limitations are set on the particle of a box problem. The box is made finite with infinite potential energy "walls."
- *Precision*—number of significant digits for an attribute measurement.
- *Procedural abstraction*—association of a procedure with a name. The procedure can then be run by using its name.
- *Procedure*—algorithm written in a computer programming language that can be run on a computer. (Also called a program or subprogram.)
- *Projection*—coordinate transformation from three-dimensions on a "globe" to a two-dimensional representation on "paper" in which distortion results in one or more properties: area, distance, shape, and/or direction.
- *Prosthetic group*—small molecule that is either noncovalently or covalently bonded to a protein to fulfill a special function.
- *Protein*—group of organic compounds composed of one or more chains of amino acids and forms an essential part of all living organisms. Typical proteins consist of 50–1000 amino acids. Humans make at least 50,000 different proteins, and a typical cell may contain 7000–10,000 different proteins.

(Q)

- *Quantum mechanics*—mathematical system used to model the structure and behavior of atoms and molecules. Predicts that there are discrete energies allowed for subatomic particles.
- *Query*—extraction of information from a database or table based on defined attribute criteria or spatial criteria.
- *Query sequence*—The sequence a researcher has and wants to match to other known sequences in a database, to determine if the sequence of interest (the query sequence) is similar to an already-described protein or gene

(R)

- *Raster graphics*—representation of images as rectangular grids of pixels. (Also called a *bitmap*.)
- *Raster model*—representation of spatial data as a grid of cells (pixels). Each cell has one numeric value.
- *Rates*—speed (or change per unit time). In Chemistry, the rate of a reaction is usually measured as a change in concentration of a reactant or product per unit time elapsed.
- *Reaction*—a process that changes a chemical substance (or substances) into another (others).
- *Regression*—a mathematical process of comparing and systematically reducing the differences between an estimated value and a known value.
- *Relation*—table in a database.

- *Reliability*—ability to rely or depend on results from a simulation to faithfully represent a real world phenomenon or system. Problems include (1) incorrect model, (2) incorrect model representation, and (3) computational errors.
- *Resolution*—rectangular grid dimensions of pixels on a computer display (e.g. 1280 × 1024); or the finest, discernible difference in reported measured values, due to electronics and physical limitations of the sensor technology

(S)

- *Sampling frequency*—number of measurements that will be reported during a given amount of time (e.g. 5 samples per second). Related to *sampling interval.*
- *Sampling interval*—amount of time between reported sampling values. Related to *sampling frequency.*
- *Sampling length*—amount of measurement time.
- *Self-defining*—information that contains a description of its structure within itself.
- *Sensitivity analysis*—an analysis performed on a numerical model to assess which parameters the model results are most sensitive to changes in.
- *Sequence homology/percent identity*—The similarity of two gene or protein sequences (often expressed as % that is the same or % identity)
- *Simulated annealing*—optimization technique/method based on an analogy of how metals that are slowly cooled end up with a *better* (lower energy, more stable) crystal state.
- *Simulation*—computer model of the behavior of a real world phenomenon or system.
- *Splines*—a set of cubic equations used to fit a fix number of points exactly
- *Solvent*—portion of a mixture that is in greater amount. A compound of interest is often dissolved in a solvent.
- *SQL*—Structured Query Language. Format used for selection from databases based on attribute data of the SELECT FROM WHERE form.
- *Steady-state*—a system that does not change with time.
- *Substitution matrix*—A matrix of scores to compare the probability of substitution of one amino acid for a different amino in differing protein sequences, incorporating empirical data from changes in similar proteins. (i.e.—What is the likelihood that the amino acid difference is due to evolutionary change in the genetic sequences)
- *Substrate*—molecule upon which a protein acts.
- *Sum of squares*—the squaring of each number and summing of the resulting set of numbers

(T)

- *Table*—two-dimensional collection of information organized by fields (columns) and records (rows), in which attributes about spatial features are stored.
- *Target sequence*—The sequence from a database that matches (with some predefined level of % identity) the query sequence

- *Time complexity function*—mathematical function that gives the time for a procedure to run given the input size of the problem.
- *Transient*—a system that does change with time.
- *Tree*—graph in which each node (except the *root*) can have only one parent but multiple children. The root has no parents.
- *Trial*-and-error—a process of trying solutions until the best one (least acceptable error) is found.

(U)

- *Uncertainty analysis*—an analysis performed with a numerical model to assess the uncertainty of model predictions resulting from the uncertainty of model parameters.
- *Unimolecular*— type of reaction initiated when one molecule undergoes a change.

(V)

- *Validation*—subjective analysis to determine if the simulation model correctly describes the real world phenomenon or system.
- *Vector graphics*—representation of images as collections of mathematical equations for lines (i.e. vectors). Often a more compact representation that allows faithful image scaling.
- *Vector model*—representation of spatial objects as points, lines (polylines), or polygons using a series of x y locations.
- *Verification*—objective analysis to determine if the computer simulation correctly implements the model.
- *Visualization*—act of creating images, diagrams, or animations to improve communication of results from a simulation model.

(W)

- *Wave function*—amplitude of a wave. In quantum mechanics, it acts as the "path" that an electron moves around the nucleus of an atom.
- *Word*—the number of bits used by a computer for its standard size, typically 32 or 64 bits.

Index

K. Brewer, C. Bareiss, *Concise Guide to Computing Foundations*,
DOI 10.1007/978-3-319-29954-9

F

G

H

I

J

K

L

M

N

O

P